P9-ARB-186

Rewriting EXTINCTION presents

THE MOST IMPORTANT COMIC BOOK ON EARTH

Stories to save the world

An anthology dedicated to saving as many species from extinction as humanly possible.

Founder, editor, and art director
Paul Goodenough

EDITORIAL Will Dennis, Tyler Jennes, Sarah Florence Lord, Bernardo Brice, Michael Perlman, Bis Stringer-Horne

CHARITY LIAISON Marie Negus

EXPERT LIAISON Jamie Woolley

DESIGNERS Cornelia Abfalter, Antonio Valjean

LEAD LETTERER Bernardo Brice

COVER ILLUSTRATION Jeff Langevin

CHAPTER ART Dan Shearn (Chapters 1, 2), Jeff Langevin (Chapter 3), Jock (Chapter 4), Tula Lotay (with thanks)

INTERVIEWERS Paul Goodenough, Sarah Florence Lord, Tyler Jennes, Bis Stringer-Horne, Alex Segura, Maddy Diment

Special thanks to:

Hannah Williams, Media Bounty, Climate 2025, Aerian, Ryan Cheetham, Ben Poole, Chris Bingham, Livia Firth, Robert Ferrell, Dickie Chappell, Veronica Treit, David Baillie, Dan Shearn, Emma Price, Jamie Kelsey, Scott Bryan Wilson, and you.

CONTENTS

DEDICATED TO...

Greenpeace

Fernando Pereira, Photographer (who was killed when the Rainbow Warrior *was bombed and sunk in 1985)*

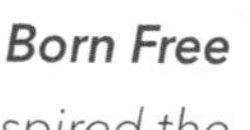

Born Free

Pole Pole's death at London Zoo inspired the creation of Born Free. She is symbolic of all wild animals in need, whether free or living in captivity, that we have dedicated our lives to helping.

Rewilding Europe

Rory Young, Conservationist/Wildlife Preservationist

Reserva

Finley Broaddus, Environmental Advocate

Stop Ecocide International

Polly Higgins, lawyer and co-founder

World Land Trust

Dedicated to all the people who are still trying to make a difference.

The Editors

Dedicated to all the species who lost their fight with extinction, and to those fighting for survival right now.

YOU ARE
AWESOME

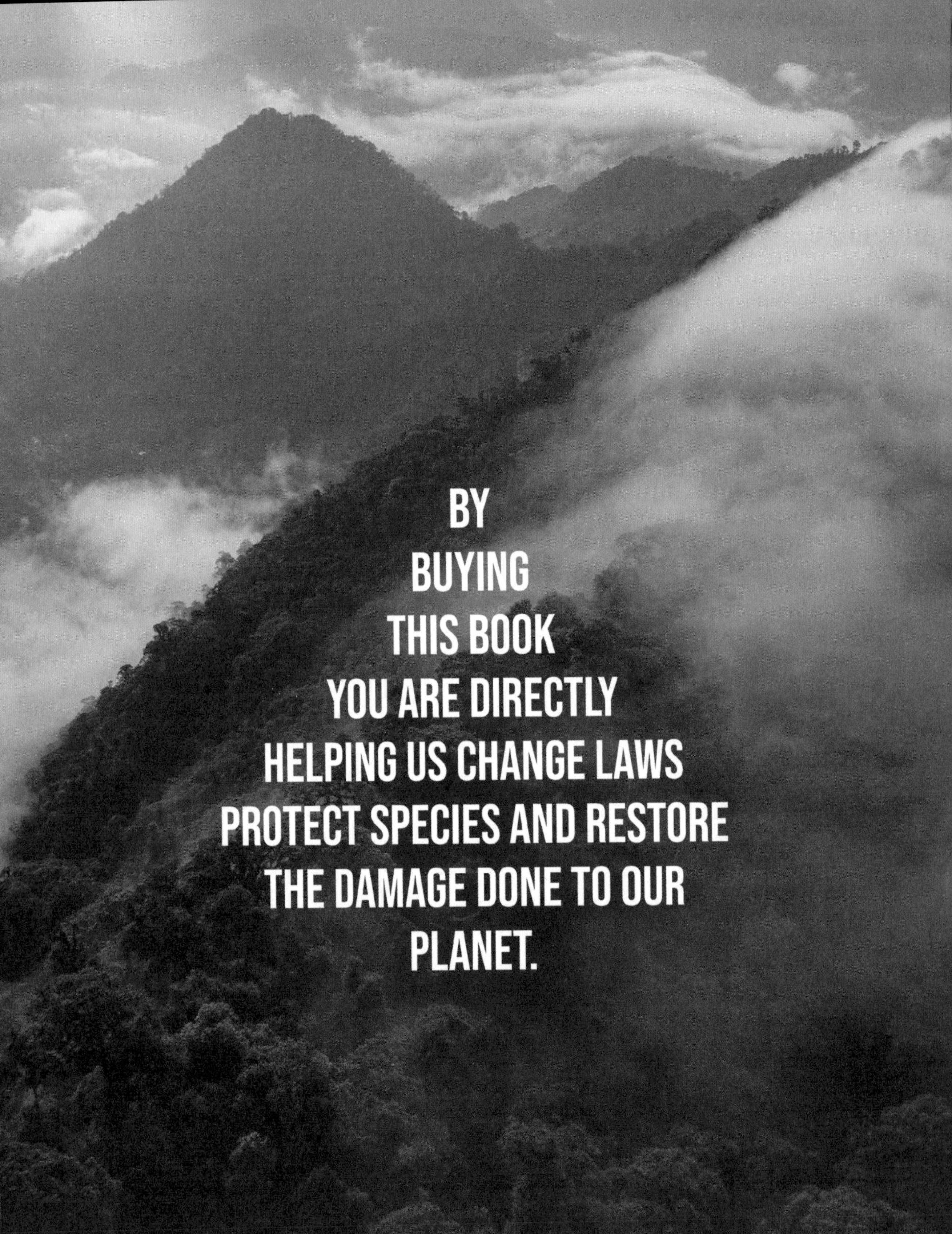

BY
BUYING
THIS BOOK
YOU ARE DIRECTLY
HELPING US CHANGE LAWS
PROTECT SPECIES AND RESTORE
THE DAMAGE DONE TO OUR
PLANET.

BUT NONE OF THIS IS POSSIBLE WITHOUT YOU.
WITHOUT YOUR DEDICATION AND ACTION.

YOU ARE NOT VOICELESS.
YOU ARE NOT POWERLESS.
YOU CAN HELP.

AND IF YOU'D LET US
WE'D LOVE TO SHOW YOU HOW...

FOREWORDS

SCOTT SNYDER

WRITER,
BATMAN, UNDISCOVERED COUNTRY, NOCTERRA

Comics. What are they? To most people they're books about caped heroes, individuals with extraordinary powers who fight universe-shattering battles. And comics are about these folks, yes, but like their titular superheroes, beneath the tights and masks, comics have a secret identity, a true face. See, deep down, at the most fundamental level, comics are about connection.

Every comic is an act of expression, an attempt to reach across distances—some literal, some existential, all real—to connect with someone else in wonder, in fear, in hope. The very act of creating them is connective. They're collaborative by nature. To make one requires sharing your deepest aspirations and anxieties with all these other people—creative partners, editors…it's a collective creation. And comic books like THIS are hardest of all to make, but also the most rewarding. So many great talents coming together to connect with you about an issue that affects us all—our ecological crisis and the urgent, dire need for ACTION.

So thank you for picking this book up. Because it is a true comic. One that inverts the old formula, maybe, but in the best way: instead of luring you in with aliens and lasers, it shows you what it's about first. Then asks YOU to don the tights, the capes, and go out and be superheroes. To get out there and change it for the better.

PAUL GOODENOUGH

FOUNDER

One day soon, these words will be obsolete...

...people will screw their face up at the idea that at one point in history, which is now, of course, there was once two hundred species going extinct every day.

They'll ask us, "What did you do during the climate crisis? What did you do to help?"

Well, we made this. Three hundred of the most passionate, astounding, and beautiful people on the planet got together and devoted their hearts and lives to making a collection of stories that could save as many species from extinction as humanly possible.

We made this book for every one of you.

We made it to help you help them–the species who have no voice.* We hope our stories open your hearts and your minds and give you the direction and power you need to join us in this fight.

Because if they do, today's problems, and every word I've written, will be obsolete. And we'll all be able to say we helped save the planet.

And what a wonderful thing that will be.

**Correct at time of writing. Check out Peter Gabriel's Interspecies Internet if you want to see how him and some amazing people are helping us (quite literally) talk to the animals.*

CHAPTER ONE

CHANGE

THE SYSTEM

Humanity is starting to realize that the system we've built cannot go on. We cannot have infinite growth on a finite planet. The following stories highlight some of the current global systems and practices that are sending us spiraling into oblivion.

With the help of organizations like Greenpeace, Stop Ecocide International, along with experts and other charities, our aim with this project is to highlight, challenge, and rewrite the system, forever.

Sponsored project:

GREENPEACE

We all see the same problems.
The world is spiraling out of our control. It's got too big. Moving too quickly.
MAKE AMERICA GREAT AGA
MAKE AMER GREAT AGA
Today, unlike generations before us, we are a global community. Today, our friends and neighbors aren't defined by locality. The problems we face aren't contained to our families or our community; they're global... continual... clawing at us every second of the day.
We are all everywhere and nowhere all at the same time. Empowered and yet powerless.
THIS TIMES
WORLD DISASTER
RIOT
STOP THIS
And so we're simply left with emotions: Love. And Rage.
Sometimes our emotions give rise to positive action, sometimes negative. But it's our reactions that divide us--in our hearts, we all want the same thing--a safe and healthy planet. And a future we can be proud of.
STOP

BREAKING NEWS
We are in the biggest and possibly most important period in human history. The actions we take now will forever rewrite the world.
We have the chance, all of us, to make the world better forever. To re-establish society as we want it.
We are a family. And you may disagree with the methods some employ.
But we are fighting to create the world everyone wants.
PEACE
A world where everyone's voice matters...

A world that really takes back control for people. That really makes their country great again.

ZERO WASTE

RECYCLE♡

SCHOOL

A world that uses inspiration and human ingenuity to make things better...

That allows us to eat well and be healthy without making others suffer...

That instills some of the wisdom of the generations before us...

That puts community and home-grown back at the center of our communities...

MAKE EARTH GREAT AGAIN

WRITTEN BY
PAUL GOODENOUGH
ART BY
SERG ACUÑA
LETTERING BY
BERNARDO BRICE

RELATIONSHIPS CAN CHANGE

DEVELOPED WITH
TOM MUSTILL
WRITTEN BY
PAUL GOODENOUGH
ART & LETTERING BY
SARAH GRALEY

It was a normal morning. I'm a composer. It's a job, believe it or not. I'm ***meant*** to be working on a score...
...but I wasn't doing anything until I had more coffee inside me and did the morning scroll.
My feed did what it promises, what it's designed to do.
I was simultaneously sated and left hungry.

A friend had posted. She lived on the other side of the world.
I did the strangest thing.
I paid attention.

I hadn't seen her for ages. She may as well not exist, really...
Sometimes it's hard to believe we all look up at the same sky.

She shared some lyrics she'd written...

They were heartfelt, lonely, wanting. The light hit the image ***just so***. It made me presume she'd used a filter.

But no. It was real.

She said it's an actual song. There's a melody and everything.

She hadn't posted the actual score.

I had no idea how the song goes...

...but she assured us all there is a melody.

I had to use my imagination.

I tried, thinking of distant people, and how we connect and...

Well, using your imagination is an act of seeing.

And, in that moment, I started to see ***more***.

And a train of thought provoked by a distant friend, an unheard melody reached an unexpected destination...

I saw notes.

No note played in the same space is alone. They ***interact***. That's what melody is.

I thought of the world as melody.

And if there's a connected melody, and the world is an orchestra playing it, it means other things.

Notes can be discordant.

A single note can harm that harmony.

And we can all question what power we truly have... but at the least, you have the power to not play discordant notes.

And encourage others to do likewise.

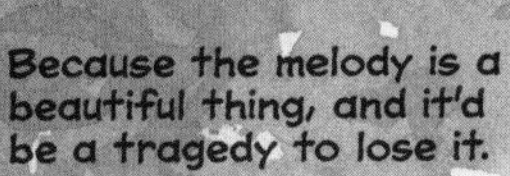

Because the melody is a beautiful thing, and it'd be a tragedy to lose it.

I decided I'd try to sing in tune with this melody. I'd find the best place for me to play in this orchestra...
That was it.
Nothing dramatic. Nothing super heroic. Just a decision. I decided and I did.

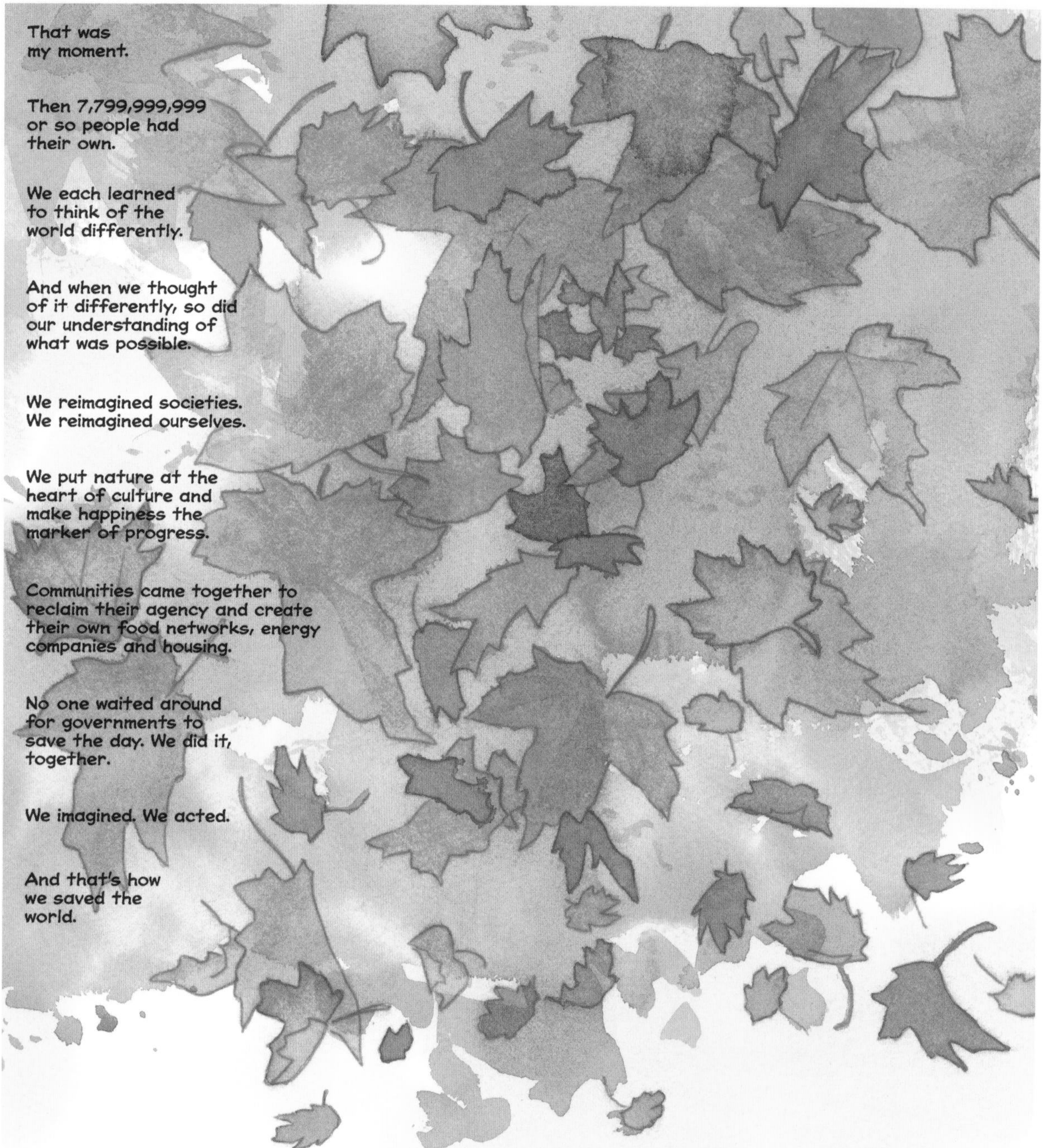

That was
my moment.

Then 7,799,999,999
or so people had
their own.

We each learned
to think of the
world differently.

And when we thought
of it differently, so did
our understanding of
what was possible.

We reimagined societies.
We reimagined ourselves.

We put nature at the
heart of culture and
make happiness the
marker of progress.

Communities came together to
reclaim their agency and create
their own food networks, energy
companies and housing.

No one waited around
for governments to
save the day. We did it,
together.

We imagined. We acted.

And that's how
we saved the
world.

Melody

DEVELOPED WITH
CARA DELEVINGNE AND ECORESOLUTION
WRITTEN BY
KIERON GILLEN
ART & LETTERING BY
SEAN PHILLIPS

THE BUNKER

DEVELOPED WITH CARA DELEVINGNE AND ECORESOLUTION | WRITTEN BY PAUL GOODENOUGH & DAMI LEE | ART & LETTERING BY DAMI LEE

MY FRIENDS... EARTH IS UNINHABITABLE. IT'S DEFINITELY NOT OUR FAULT THOUGH.
EARTH = CANCELED?

WHICH IS WHY I'VE CREATED A BUNKER FOR US TO HIDE IN FOR SAFETY! JOIN ME IN THE BUNKER UNTIL ALL OF THIS BLOWS OVER!
YEAH!
HEAR, HEAR!
CORPORATE INTERESTS
FACTORY FARMING
POLITICIAN
I ♥ FOSSIL FUELS

YES, STEP RIGHT IN!
DA BUNKER
IN YOU GO!

SLAM

WHOOSH
DAMI

NOW THAT WE'VE GOT THEM OUT OF THE WAY, IT'S TIME FOR THE REAL WORK TO BEGIN!
PEOPLE POWER

SMOTHER EARTH

DEVELOPED WITH JOJO MEHTA

WRITTEN & DRAWN BY HANNAH HILLAM

DID YOU KNOW THAT IN MOST OF THE WORLD, DESTRUCTION OF NATURE ISN'T A CRIME?

By Jojo Mehta

WHAT DOES THAT MEAN?

Well, most laws about how we treat the natural living world are in the form of environmental regulations, lists of detailed "do's and don'ts." Big companies pay expensive lawyers to work around these rules, and often get little more than a fine and a slap on the wrist if they don't follow them.

What we need to do is make destroying the planet an international crime.

This means the people who make the decisions can't hide behind their businesses, but can be brought to justice individually and held personally responsible.

Chief executives, board members, and key decision makers in the world's most destructive companies could face jail time if their decisions lead directly to serious planetary damage.

Imagine what a difference this would make...

WHY IS IT SO IMPORTANT?

Protecting the future of life on Earth means stopping the mass damage and destruction of ecosystems taking place globally.

We call this serious harm to nature Ecocide. And right now, in most of the world, no-one is held responsible for it.

It's time to change the rules.

It's time to make Ecocide an international crime.

Together with lawyers, diplomats, and civil society, we* work toward amending the Rome Statute of the International Criminal Court to include a crime of Ecocide.

The Rome Statute governs the "most serious crimes of concern to the international community as a whole," currently Genocide, Crimes Against Humanity, War Crimes, and the Crime of Aggression.

By adding Ecocide to this list, we send a powerful message to the whole world that severe harm to nature is both legally and morally unacceptable.

WHAT DOES THAT MEAN?

Follow **Stop Ecocide International** for more information and actions you can take.

*Stop Ecocide International (global campaign) and the Stop Ecocide Foundation (NL charitable foundation)

ECOCIDE

DEVELOPED WITH JOJO MEHTA | WRITTEN BY PAUL GOODENOUGH & HANNAH HILLAM | ART & LETTERING BY HANNAH HILLAM

THE XR DIARIES

WRITTEN BY
JOHN WAGNER
ART BY
LEONARDO MARCELLO GRASSI
LETTERING BY
BERNARDO BRICE

LIAM NORTON, ELECTRICIAN. U.K.

I HEARD ABOUT EXTINCTION REBELLION FROM THE LETTER IN THE GUARDIAN AND WENT TO THE BRIDGE BLOCKADE, JUST TO SEE WHAT WAS GOING ON REALLY.

FOR MY CHILDREN STOP! CLIMATE C
FOR ALL LIFE ON EARTH WAKE UP GOVERN ACTION
BEL FOR LIF

IT GOT ME INTERESTED ENOUGH TO ATTEND A "HEADING FOR EXTINCTION" MEETING. UNTIL THEN, I SUPPOSE, I'D BEEN IN DENIAL. OH, I KNEW THERE WAS A PROBLEM WITH THE CLIMATE; I JUST HADN'T REALIZED HOW BAD THINGS WERE. OR PERHAPS I DIDN'T WANT TO KNOW.

I REMEMBER SAYING TO ONE OF THE PEOPLE LEADING THE MEETING--"IF EVERYTHING YOU'VE SAID IS TRUE THEN OUR JOBS DON'T MATTER."

A COUPLE OF WEEKS LATER I WAS THROWING FAKE BLOOD OUTSIDE DOWNING STREET.

"I'd been in denial."

I SEE GETTING ARRESTED AS A NECESSARY SACRIFICE, ANOTHER WAY OF HELPING TO FORCE CHANGE. BECAUSE THE GREAT MOMENTS IN HISTORY HAVE ONLY COME ABOUT BECAUSE PEOPLE WERE WILLING TO STAND UP FOR WHAT THEY BELIEVED IN. TO **DO** SOMETHING.

IF THERE'S ONE THING I'D SAY ABOUT GETTING ARRESTED, IT'S DON'T BE INTIMIDATED BY IT. BEING ARRESTED MEANS YOUR ACTIONS RIPPLE THROUGH THE LEGAL SYSTEM-- WHICH IS EXPENSIVE AND DISRUPTING FOR THEM, AND FORCES THEM TO PAY ATTENTION.
ALSO, AS A WHITE MAN, I'M IN A PRIVILEGED POSITION-- THE SYSTEMATIC INJUSTICES MAKE IT MUCH HARDER FOR PEOPLE OF COLOR TO GET INVOLVED WITH THESE KINDS OF ACTIONS. THE MORE OF US THAT DISRUPT THE SYSTEM, THE MORE THE SYSTEM WILL BE FORCED TO CHANGE.

I'M CONVINCED THE JUDGE WAS SWAYED BY OUR ARGUMENTS, ENOUGH TO FIND A TECHNICALITY TO GIVE US A NOT GUILTY VERDICT. HER OWN KIND OF REBELLION. EVEN PEOPLE AT THE TOP OF THE SYSTEM CAN SEE THE DANGER OUR PLANET IS IN.
ACT NOW
SINCE THEN I'VE TAKEN PART IN LOADS MORE ACTIONS. I'M GETTING QUITE USED TO BEING ARRESTED ACTUALLY. ONCE YOU'VE REALIZED WHAT A CATASTROPHE COULD BE COMING OUR WAY, BEING ARRESTED IS NOTHING.

I VOLUNTEERED TO HELP THE RELIEF MISSION IN LAMU DURING THE TERRIBLE DROUGHT IN 2017. WHAT I SAW BROKE MY HEART.
FAZEELA MUBARAK, ACCOUNTANT. KENYA.

THE RAINS HAD FAILED, AND TEMPERATURES WERE UNUSUALLY HIGH. DESOLATION AND DESPERATION WERE EVERYWHERE, CORPSES OF ANIMALS SCATTERED ON THE GROUND.

"People are right now dying from the effects of climate change."

ANIMALS THAT STILL SURVIVED WERE SEARCHING DESPERATELY FOR WATER. MANY COVERED THEMSELVES IN MUD AS PROTECTION FROM THE SUN. SOME, THAT VENTURED TOO FAR INTO THE MUDHOLE, NEVER ESCAPED.

FOR THOUSANDS OF YEARS THE PEOPLE HAVE LIVED IN HARMONY WITH ANIMALS, BUT IT IS A DELICATE ECOSYSTEM, EASILY UPSET, AND PROLONGED DRY SEASONS CAUSED BY CLIMATE CHANGE PUT IT UNDER A TERRIBLE STRAIN.

PEOPLE WERE SUFFERING, BUT STILL THEY ASKED US TO HELP THE ANIMALS FIRST. ANIMALS DESPERATE FOR FOOD AND WATER CAN TRAMPLE FARMS AND DESTROY VILLAGES.

PEOPLE NEED HELP. THEY NEED HELP DIGGING WATER HOLES, HELP TO PROTECT THEM AGAINST DROUGHT AND FLOOD.

GIVE THEM FRESH WATER AND THEY CAN HELP TO LOOK AFTER THE ANIMALS TOO.

IT IS A START, BUT IT IS NOT THE WHOLE SOLUTION. THEY TELL US THE WORLD IS GOING TO CONTINUE TO GET HOTTER. WILL PEOPLE AND ANIMALS STILL BE ABLE TO SURVIVE IN AFRICA? I PRAY THAT WE WILL.

BUT TO MAKE THAT POSSIBLE WE MUST HAVE FEWER EMPTY PROMISES AND MORE ACTION. AND TIME IS RUNNING OUT.

CATHY EASTBURN,
MUSICIAN.
U.K.
I WAS AWARE THAT PEOPLE WERE CONSUMING TOO MUCH, BUT I HOPED WE'D BE OK AND REIGN OURSELVES IN. YET FOR THE LAST 20 YEARS I'VE HAD THE GROWING FEAR THAT WE WEREN'T GOING IN THE RIGHT DIRECTION.
EX

WHEN THE IPCC* REPORT ON CLIMATE CHANGE CAME OUT IN OCTOBER 2018 IT TERRIFIED ME. TWELVE YEARS, IT SAID--WE HAD JUST **TWELVE YEARS** TO AVERT A CLIMATE CATASTROPHE.
* U.N. INTERGOVERNMENTAL PANEL ON CLIMATE CHANGE.

MY GENERATION KNEW ABOUT THIS STUFF LONG AGO, AND WE ALL PRETENDED IF WE JUST DID THE SMALL THINGS--A BIT OF RECYCLING, LESS FLYING--EVERYTHING WOULD BE ALL RIGHT. BUT IT WASN'T GOING TO BE ALL RIGHT.
I FELT ASHAMED, AND TERRIFIED FOR MY CHILDREN. I REMEMBER HUGGING MY DAUGHTER AND FEELING RESPONSIBLE FOR THIS WHOLE MESS.

"I need to take action for my children. Or else my children won't have a future."

I'M NOT BY NATURE A DISRUPTIVE PERSON, BUT I DIDN'T FEEL I HAD A CHOICE. I COULDN'T JUST STAND BY. WITH OTHERS IN EXTINCTION REBELLION I BEGAN TO TAKE ACTION. NINE OF US GLUED OURSELVES TO HOTEL RAILINGS.

MEANWHILE, LIAM FOX WAS SPEAKING, DISCUSSING GOVERNMENT PLANS TO INCREASE INVESTMENT IN THE INTERNATIONAL OIL INDUSTRY. PEOPLE IN POWER WEREN'T LISTENING. WE HAD TO MAKE THEM LISTEN.

THE DAY WE DEMONSTRATED ABOARD A DLR TRAIN AT CANARY WHARF I'D BEEN SICK WITH FEAR. I'M A MUSICIAN, NOT AN URBAN WARRIOR.
EMERGENCY

ary Wha
or Jubilee line
I DON'T RELISH MAKING LIFE DIFFICULT FOR PEOPLE, BUT UNLESS YOU CAUSE SOME DISRUPTION NOBODY NOTICES.
AFTER CANARY WHARF THREE OF US WERE ARRESTED. IN COURT WE WERE REFUSED BAIL AND TAKEN STRAIGHT TO PRISON.
IT WAS QUITE A SHOCK, THOUGH I REALIZE IF WE BREAK THE LAW WE HAVE TO BE PREPARED TO TAKE THE CONSEQUENCES.

IN JAIL I WATCHED THE DEMONSTRATION UNFOLD ON TV. I FELT PROUD AND PLEASED. WE WERE MAKING OUR POINT.

AT OUR TRIAL WE PLEADED THAT OUR ACTIONS WERE TAKEN IN SELF-DEFENSE. WE WERE TRYING TO SAVE OUR LIVES-- AND THE PLANET.

THE JUDGE INSTRUCTED THE JURY THEY HAD TO FIND US GUILTY OR NOT GUILTY BASED ONLY ON THE CIRCUMSTANCES OF THE ARREST, BUT THE JURY CONDUCTED THEIR OWN LITTLE REBELLION. THEY FOUND US GUILTY--'WITH REGRET.'

I'VE BEEN ARRESTED FIVE TIMES NOW, AND COUNTING. TOO MANY PEOPLE ARE STILL BEHAVING AS IF IT'LL BE FINE, AND IT WON'T.
WE ONLY HAVE ONE PLANET, AND I NEED TO PRESERVE IT FOR MY CHILDREN--FOR YOUR CHILDREN--AND FOR THE GENERATIONS TO COME. WE OWE IT TO THEM.

DETECTIVE SERGEANT PAUL STEPHENS. U.K.

I JOINED THE METROPOLITAN POLICE IN 1983 AND RETIRED IN 2018 WITH THE RANK OF DETECTIVE SERGEANT.

MOST OF MY SERVICE WAS IN CRIMINAL INVESTIGATION, INTELLIGENCE, HOMICIDE, HATE CRIME, RISK MANAGEMENT AND PREVENTION.
EVA, MY WIFE, IS SWISS. WE'VE SPENT MANY YEARS TAKING HOLIDAYS IN THE SWISS MOUNTAINS, AND EACH YEAR WE'VE NOTICED THE GLACIERS GROWING SMALLER AND MORE HOLES IN THE ICE APPEARING. EACH YEAR IT'S BEEN GETTING WORSE.

IT ENCOURAGED ME TO START RESEARCHING GLOBAL WARMING. THERE IS A WEALTH OF SCIENTIFIC EVIDENCE OUT THERE.
IT BECAME CLEAR TO ME THAT HUMANITY, OBSESSED WITH ECONOMIC GROWTH, COULD ONLY CONTINUE BY EXPLOITING MORE AND MORE RESOURCES ON A FINITE PLANET. WE WERE DOOMED.
EVA AND I RESIGNED OURSELVES TO THE FACT THAT HUMANITY DID NOT HAVE MUCH FUTURE.

"The danger is so great that it threatens the very existence of our species."

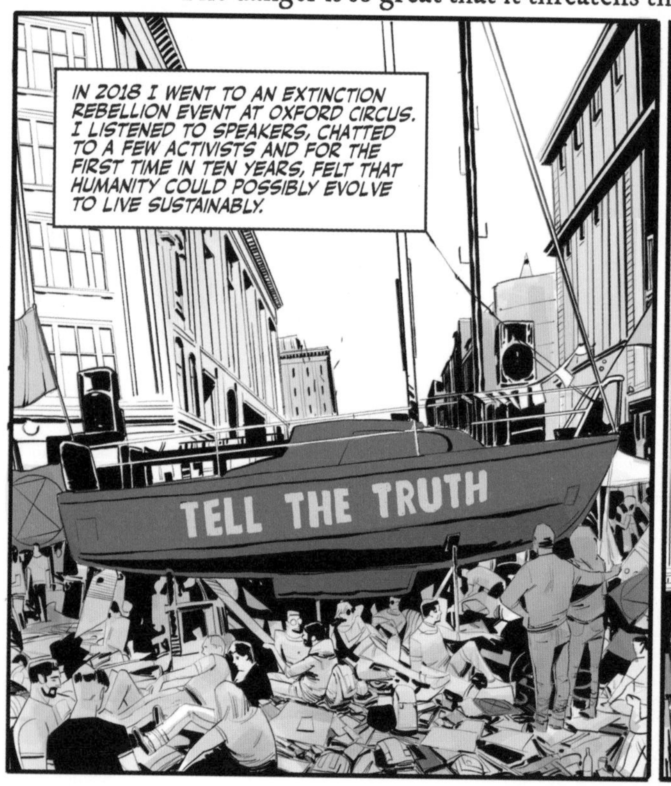

IN 2018 I WENT TO AN EXTINCTION REBELLION EVENT AT OXFORD CIRCUS. I LISTENED TO SPEAKERS, CHATTED TO A FEW ACTIVISTS AND FOR THE FIRST TIME IN TEN YEARS, FELT THAT HUMANITY COULD POSSIBLY EVOLVE TO LIVE SUSTAINABLY.
TELL THE TRUTH

AS A RETIRED COP I COULD SEE THAT THESE OTHERWISE LAW-ABIDING CITIZENS BLOCKING THE ROAD, PEACEFULLY EDUCATING PASSERSBY ABOUT THE CLIMATE CRISIS, PRESENTED A DILEMMA TO POLICE.
PERSONALLY, I HAD COME TO SEE LITTLE DIFFERENCE BETWEEN KEEPING THE PUBLIC SAFE WITH POLICING AND KEEPING THE PUBLIC SAFE BY RAISING THE ALARM ABOUT OUR CLIMATE.

IN THE END POLICE DID MOVE IN. THEY ARRESTED OVER 1,100 PEOPLE, SO MY FIRST ROLE WAS GIVING SUPPORT AT COURT. IT WAS HEARTENING TO SEE THE ATTITUDE OF THE JUDICIARY MELT FROM DISMISSIVE IRRITATION TO WHOLE-HEARTED AND EMOTIONALLY ENGAGED SUPPORT.

I THEN SOUGHT OUT POLICE LIAISON AND OFFERED MY HELP. I ATTENDED EVERY ACTION THAT I COULD AND LIAISED WITH POLICE, OFTEN WITH ANOTHER FORMER OFFICER, TO HELP KEEP THINGS AS SAFE AS POSSIBLE AND ENSURE THAT DEMONSTRATORS KNEW THE CONSEQUENCES OF THEIR ACTIONS.

IN THE OCTOBER ACTION, AFTER THE BAND MASSIVE ATTACK HAD ADDRESSED THE CROWD, I WAS DRAGGED ONTO THE PLATFORM. I RECOUNTED A CONVERSATION I HAD HAD WITH POLICE...
A LOT OF POLICE HAVE TOLD ME THAT "WE ARE BEHIND WHY YOU'RE DOING THIS, JUST NOT HOW YOU'RE DOING IT. THERE ARE PROPER DEMOCRATIC METHODS."
LOOK AROUND YOU, I TOLD THEM.

THIS IS REAL DEMOCRACY.

I'VE SINCE BEEN INTERVIEWED ON BREAKFAST TELEVISION AND ON LOCAL RADIO. FINE WITH ME. I'LL GO ON DOING WHATEVER I CAN TO HELP.
WHEN THE DANGER IS SO GREAT THAT IT THREATENS THE VERY EXISTENCE OF OUR SPECIES, AS WELL AS EVERY OTHER SPECIES THAT WE LIVE WITH AND DEPEND UPON, I FIND IT INCREDIBLE THAT APPARENTLY INTELLIGENT PEOPLE DO NOT GRASP THE FACT THAT WE HAVE TO CHANGE DRASTICALLY.

WE HAVE TO CHANGE NOW.

THE MOON.
TOMORROW.
MAN ON
THE MOON

"OCEANS OF DEEPEST BLUE...

"CREATURES OF UNTOLD COLOR AND BEAUTY...

"A WORLD ***OVERFLOWING*** WITH LIFE.

"WE HAD IT ***ALL.***"

WHEN CAN WE GO BACK?

WHAT DID YOU DO?
WRITTEN BY
BELLA LACK
ART & COLORS BY
MORGAN BEEM
LETTERING BY
JIM CAMPBELL
THE END

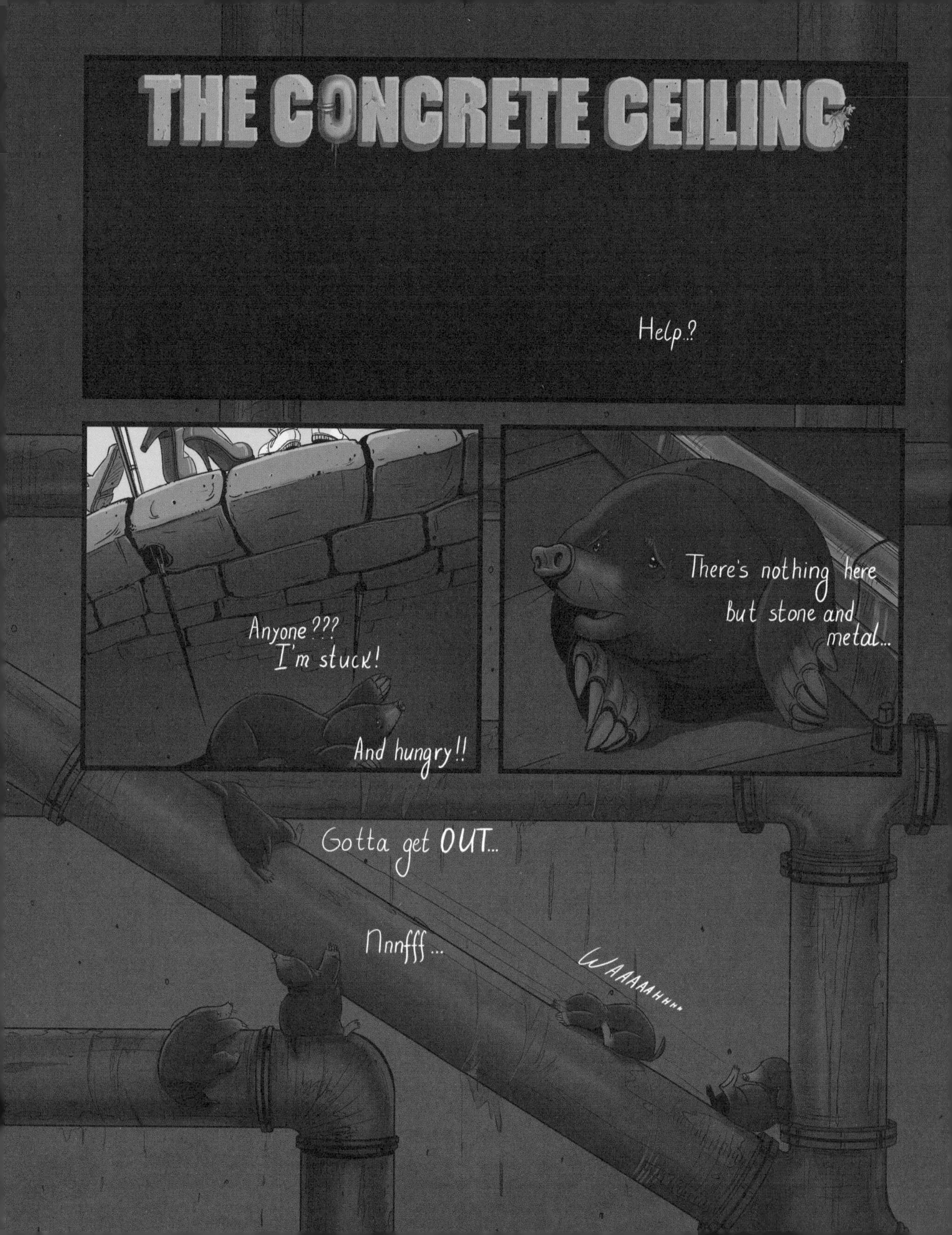

THE CONCRETE CEILING

Help..?

Anyone???
I'm stuck!

And hungry!!

There's nothing here
but stone and
metal...

Gotta get **OUT**...

Nnnfff...

WAAAAAHHHH...

HONK
HONK
HOOOONNNKKK!!!
Can't...
...get out!
Please...

A hole!
Must escape!
Food???
HONK
HONK
No...
No!
NOOOO!

What is this hell?
Can't go on...
I give up...
Sniff
sniff

...?

What are they doing?

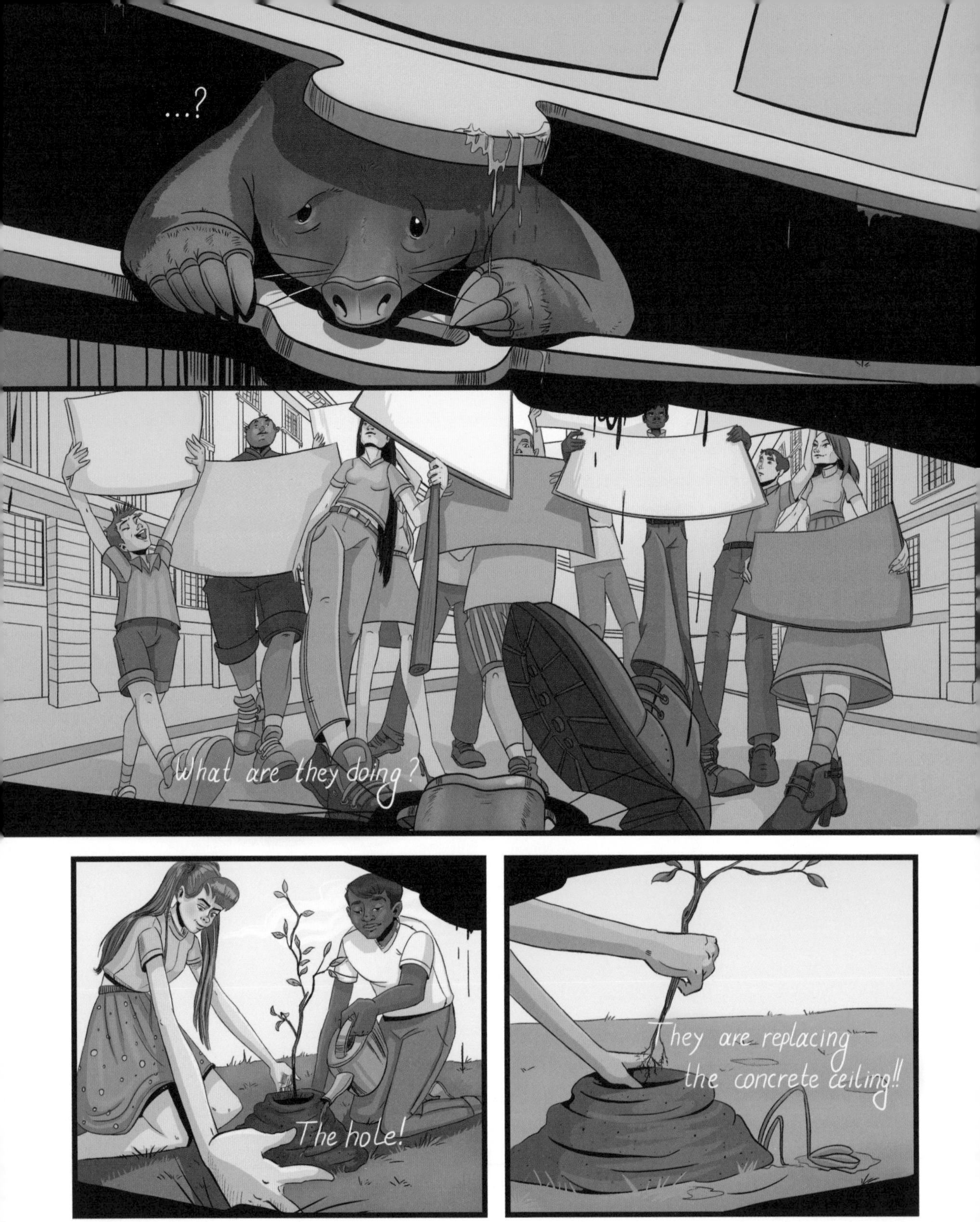

The hole!

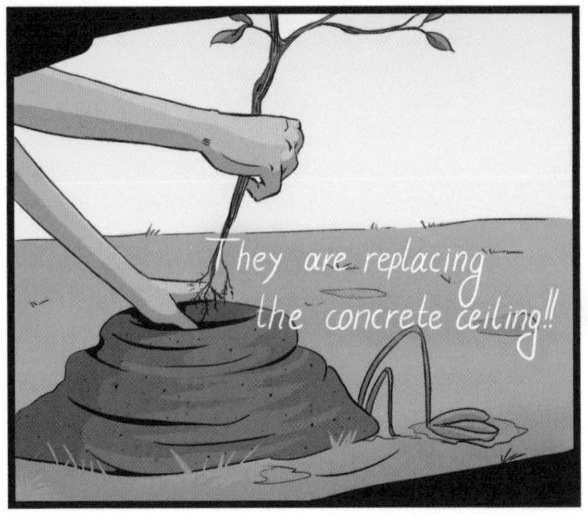

They are replacing
the concrete ceiling!!

STORY INSPIRED BY
LUISA NEUBAUER AND FRIDAYS FOR FUTURE

WRITTEN BY
PAUL GOODENOUGH

ART & LETTERING BY
ROSITSA VANGELOVA

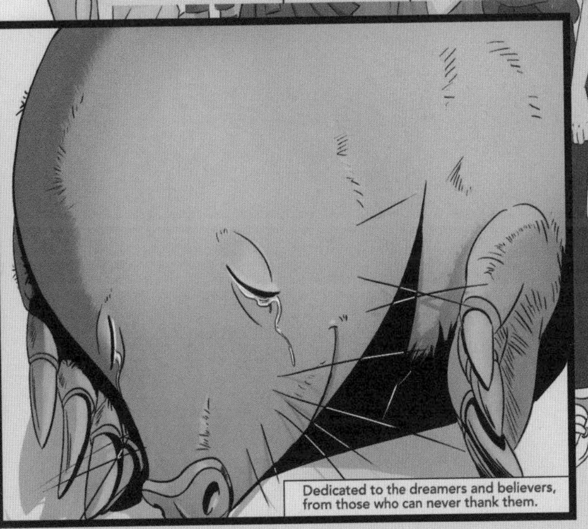

Dedicated to the dreamers and believers, from those who can never thank them.

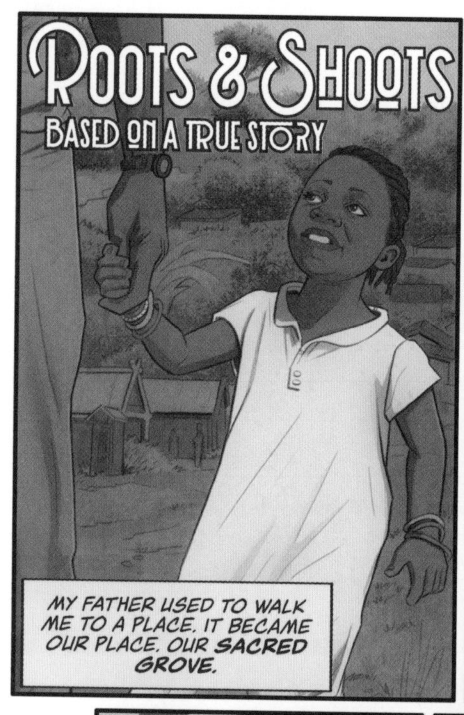
ROOTS & SHOOTS
BASED ON A TRUE STORY
MY FATHER USED TO WALK ME TO A PLACE. IT BECAME OUR PLACE. OUR SACRED GROVE.

PLACES BECOME MEANINGFUL BECAUSE TIME AND LOVE MAKE THEM SO...

CENTURIES OF LIFE CAN GIVE A PLACE A SPIRITUAL POWER...

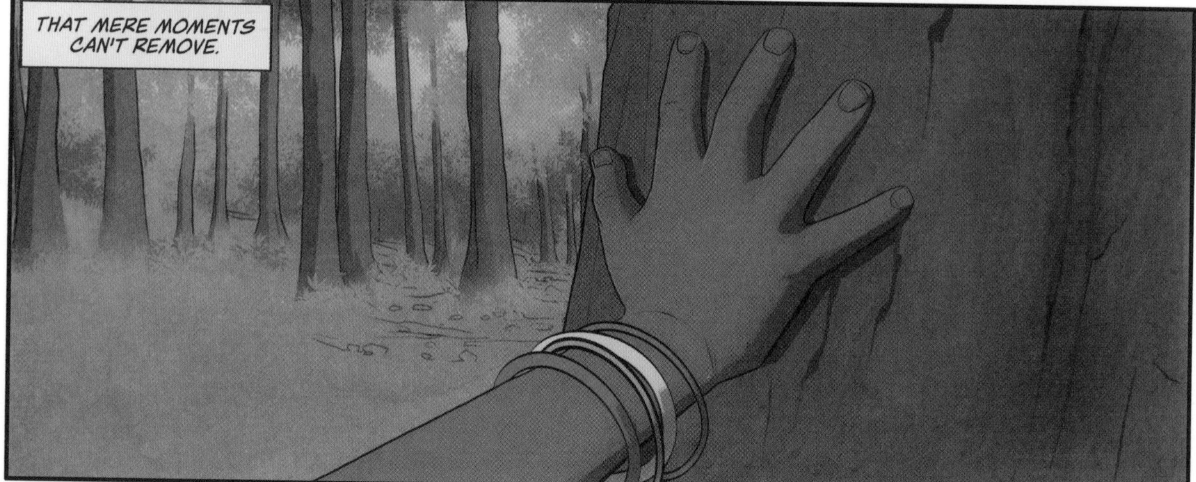
THAT MERE MOMENTS CAN'T REMOVE.

HEY! YOU DON'T BELONG HERE.

THIS IS OUR PLACE. IT DOESN'T BELONG TO YOU!

LISTEN, PLEASE, THIS IS OUR CAMP NOW. YOU CANNOT BE HERE.

I SAW YOU HERE! AS A BOY! DON'T PRETEND YOU FORGOT!

WELL?

WEEKS OF LETTER WRITING AND PETITIONING LATER...
PLEASE! WE JUST WANT TO REPLANT THE GROVE. YOU HAVE TO LET US IN.
I CANNOT LET YOU ENTER. CIVILIANS ARE NOT ALLOWED IN A MILITARY CAMP!

roots&shoots
YOU SURE ABOUT THAT?!

THEY SAY THEY WON'T BE IN THE WAY.
THEY JUST WANT TO REPLANT THE GROVE AND THEY'LL BE GONE.

THEY'VE GOT PERMISSION...
HMM.
THE CHILDREN... THEY WROTE LETTERS.

OK, FINE.
JUST THE CHILDREN, THEN.

YOU CAN'T SERIOUSLY EXPECT THEM TO--
THE LETTER SAYS THE CHILDREN HAVE BEEN GRANTED RIGHTS TO REPLANT THE GROVE. NOT YOU.

YOU CAN'T DO THIS!
THEY'RE TOO LITTLE! YOU HAVE TO LET US HELP!

THE SUN WAS SO VERY, VERY HOT THAT DAY.

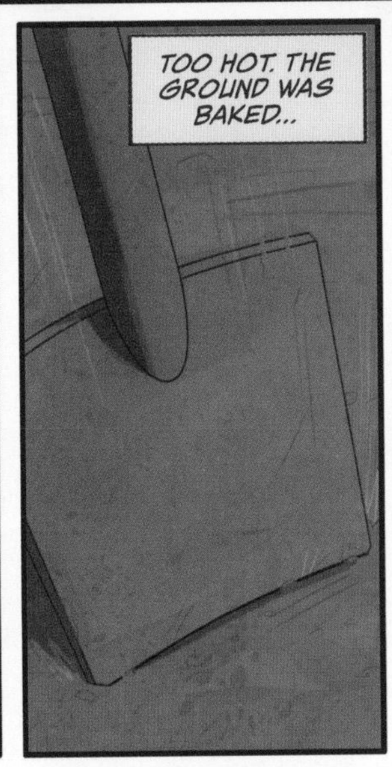
TOO HOT. THE GROUND WAS BAKED...

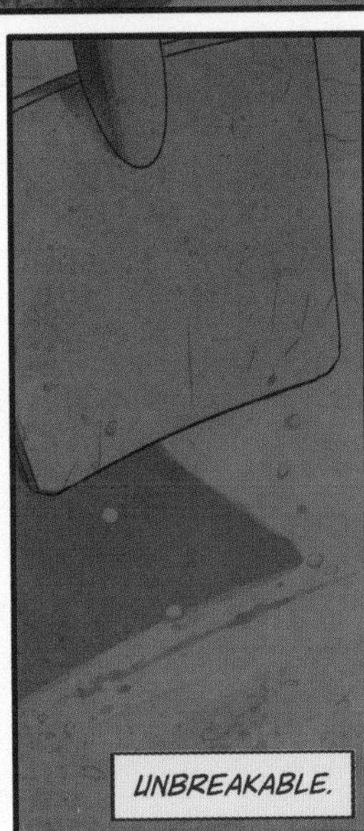
UNBREAKABLE.

WE'D COME SO FAR... DONE SO MUCH...

OUR FAMILIES URGED US ON WITH ALL THEIR HEARTS.

BUT IN THEIR EYES THEY COULD SEE WHAT WE COULD SEE...
DESPITE ALL OUR **EFFORTS**. WE JUST WEREN'T **STRONG** ENOUGH.

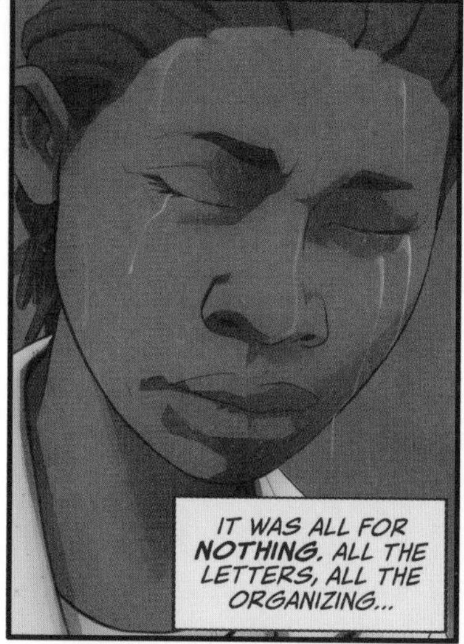
IT WAS ALL FOR **NOTHING.** ALL THE LETTERS, ALL THE ORGANIZING...

SO I COLLAPSED...UNDER THAT BAKING SUN, SURROUNDED BY THE REMAINS OF THE GROVE I FOOLISHLY THOUGHT I WOULD SAVE.

BUT THEN I HEARD SOMETHING, THE SOUND OF METAL BEING PLACED ON THE GROUND...

OF BOOTS CRUNCHING ON BAKED SOIL AS THEY APPROACHED. BUT I COULDN'T LOOK UP...I COULDN'T FACE WHAT CAME NEXT.

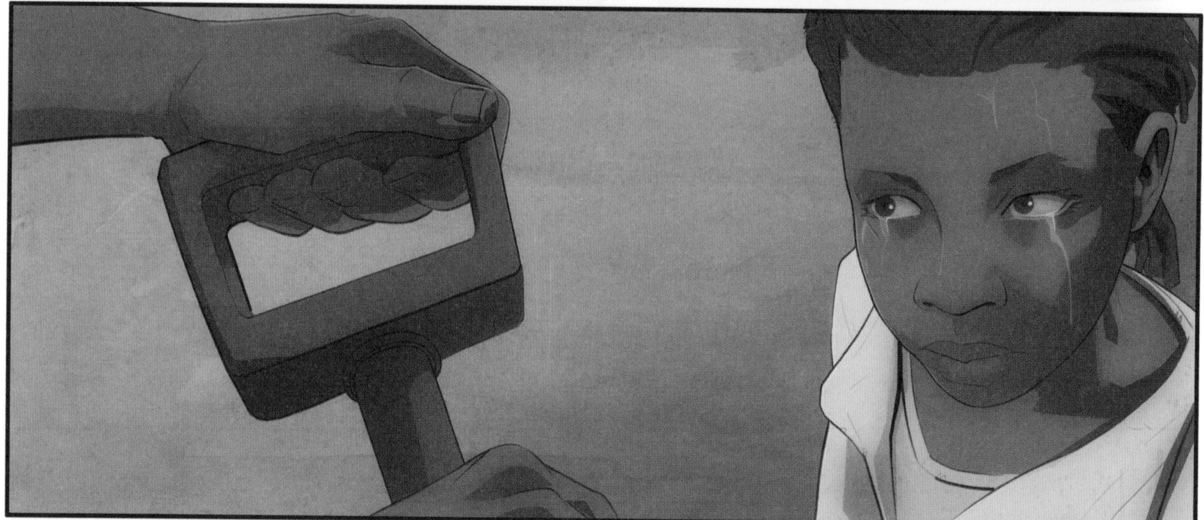

THE SOLDIERS PUT DOWN THEIR GUNS, AND PICKED UP OUR SHOVELS.
IN THAT MOMENT, WE ALL FORGOT WHO WE WERE.

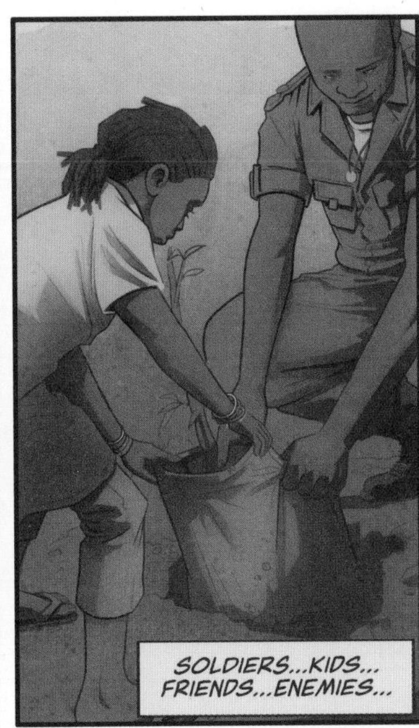
SOLDIERS...KIDS... FRIENDS...ENEMIES...

...AND WE WORKED TOGETHER TO MAKE THINGS BETTER.

AND ALTHOUGH OUR OLD GROVE HAD BEEN DESTROYED, WE MADE A NEW ONE...

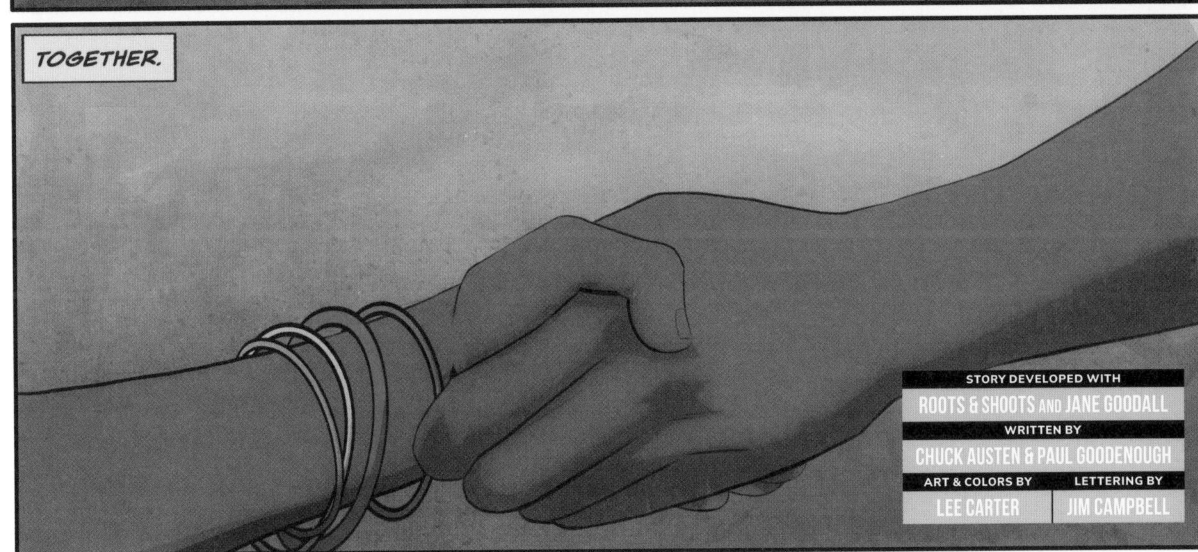
TOGETHER.
STORY DEVELOPED WITH
ROOTS & SHOOTS AND JANE GOODALL
WRITTEN BY
CHUCK AUSTEN & PAUL GOODENOUGH
ART & COLORS BY
LEE CARTER
LETTERING BY
JIM CAMPBELL

"LITTLE SUSIE WANTS SOME ICE CREAM.
"SO SHE DOES A POWER ANALYSIS: HOW CAN I GET THAT ICE CREAM?

"WELL--SHE THINKS--MOM COULD GIVE IT TO ME, BUT SHE GAVE ME ONE YESTERDAY.
"(AND SHE'S KINDA MAD AT ME ANYWAY.)

CREAM
"SO SHE GOES FOR DAD.
"BUT IF SHE JUST ASKS DAD FOR ICE CREAM, HE'LL SAY, 'GO ASK MOM.'

"SO INSTEAD, SHE HAS TO CREATE AN ATMOSPHERE WHEREBY DAD WANTS TO GIVE HER ICE CREAM.
"SO LITTLE SUSIE CRAWLS INTO HIS LAP--

"AND SAYS, 'DADDY! I CRIED MYSELF TO SLEEP LAST NIGHT, BECAUSE YOU FORGOT TO GIVE ME A HUG!
"'CAN I HAVE ICE CREAM?'

"AND SOONER OR LATER... DADDY GIVES HER SOME ICE CREAM."

Dima Litvinov is a Senior Campaigner for Greenpeace, and he explains that's all campaigning really is--
Storytelling.
Dima just happens to be telling the most important story in the world.

"WHEN I WAS TWENTY, I VISITED ECUADOR AND LIVED IN SOME VILLAGES.

"ONCE, MY LANDLORD WENT DOWN INTO THE RAINFOREST TO BUY STUFF FROM THE INDIGENOUS PEOPLE THERE.

"WE FOUND THIS BONFIRE, AND THE PEOPLE THERE WERE HUNG OVER FROM PARTYING. SO WE ALL HUNG OUT, AND I WAS IN AWE BY HOW BEAUTIFUL AND AMAZING THEY WERE.

"THE NEXT YEAR, I WENT BACK AGAIN.

"BUT THIS TIME, THERE WAS NO RIVER, NO FOREST, AND NO CAMP. A FUEL COMPANY DESTROYED IT ALL LOOKING FOR OIL. BUT THEY DIDN'T FIND ANY, SO THEY JUST MOVED ON.

"AND *THAT* WAS WHAT MOVED ME."

I WAS BORN IN MOSCOW. WHEN I WAS SIX YEARS OLD, MY FAMILY WAS EXILED TO SIBERIA, BECAUSE MY FATHER PROTESTED AGAINST THE SOVIET INVASION OF CZECHOSLOVAKIA.

SO I GREW UP WITH THE KGB SEARCHING OUR HOUSE, PEOPLE GOING INTO PRISON CAMPS, ALL THAT STUFF--

THE ***POWER*** OF AUTHORITY, BUT ALSO, THAT YOU HAVE ***ALL*** THE RIGHT TO QUESTION THAT AUTHORITY...

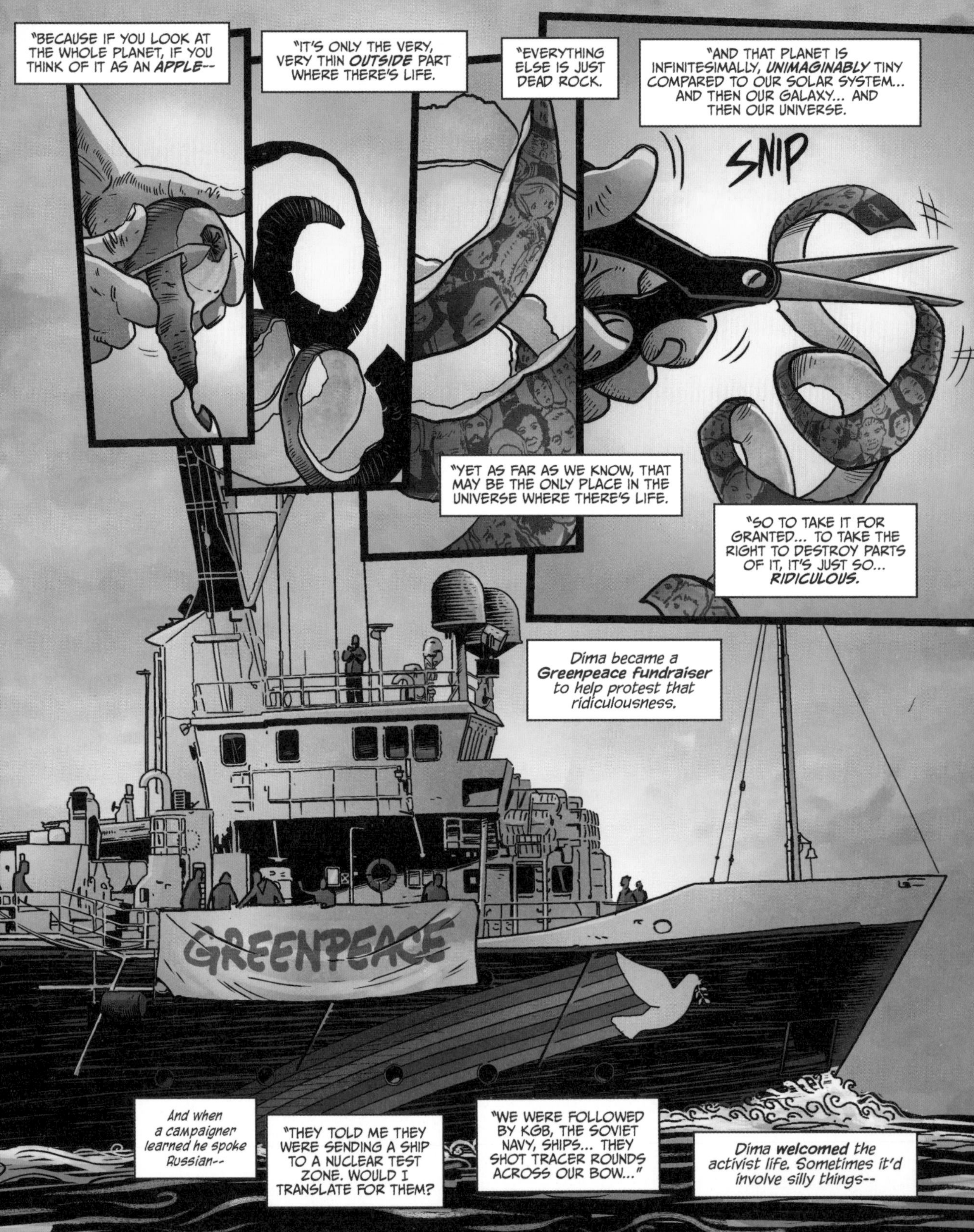
"BECAUSE IF YOU LOOK AT THE WHOLE PLANET, IF YOU THINK OF IT AS AN APPLE--
"IT'S ONLY THE VERY, VERY THIN OUTSIDE PART WHERE THERE'S LIFE.
"EVERYTHING ELSE IS JUST DEAD ROCK.
"AND THAT PLANET IS INFINITESIMALLY, UNIMAGINABLY TINY COMPARED TO OUR SOLAR SYSTEM... AND THEN OUR GALAXY... AND THEN OUR UNIVERSE.
SNIP
"YET AS FAR AS WE KNOW, THAT MAY BE THE ONLY PLACE IN THE UNIVERSE WHERE THERE'S LIFE.
"SO TO TAKE IT FOR GRANTED... TO TAKE THE RIGHT TO DESTROY PARTS OF IT, IT'S JUST SO... RIDICULOUS.
Dima became a Greenpeace fundraiser to help protest that ridiculousness.
GREENPEACE
And when a campaigner learned he spoke Russian--
"THEY TOLD ME THEY WERE SENDING A SHIP TO A NUCLEAR TEST ZONE. WOULD I TRANSLATE FOR THEM?
"WE WERE FOLLOWED BY KGB, THE SOVIET NAVY, SHIPS... THEY SHOT TRACER ROUNDS ACROSS OUR BOW..."
Dima welcomed the activist life. Sometimes it'd involve silly things--

"THE NORWEGIAN COAST GUARD WAS TRYING TO BOARD US, SO WE SURROUNDED THE SHIP WITH CHICKEN WIRE AND A BANNER SO THEY COULDN'T EASILY GET ON.
"MY JOB WAS TO STAND WITH A BOTTLE OF KETCHUP, AND WHEN THEY TRIED TO STICK A KNIFE THROUGH AND GET IN, I SPIT OUT THE KETCHUP AND YELL--"
AAHHH!
NOOO!
When the Coast Guard finally boarded, threatening to take away Greenpeace's cameras and PR, the ship's Captain replied--
THIS IS NOT ABOUT PR. IT'S ABOUT SAVING THE FUTURE OF THE PLANET!
Words Dima took to heart. Because while an organization trying to change the planet needed that PR, their actions could never be solely about getting attention. Unless it was a real protest trying to halt actual destruction, it would just be... a bad story.
ARCT
And sometimes it came with a cost.
IN 2013, DIMA AND THE REST OF THE "ARCTIC 30"-- 28 GREENPEACE ACTIVISTS AND TWO JOURNALISTS--SAILED INTO THE RUSSIAN ARCTIC TO PEACEFULLY PROTEST OIL DRILLING.
RUSSIAN AUTHORITIES RESPONDED BY DISPATCHING OFFICERS ARMED WITH GUNS AND KNIVES TO SEIZE THE SHIP.
ACTIVISTS WERE BEATEN AND JAILED FOR TWO MONTHS, SPARKING GLOBAL PROTESTS BEFORE THEY WERE FINALLY GRANTED AMNESTY...

It's one reason I wanted to speak with him. I wanted to know--when the world sometimes seems to only change for the worse, how does he stay motivated?
But listening to him, I'm instead struck by how upbeat he is. His great sense of humor.

And I'm left asking--how does he maintain that?
THE STUFF WE FIGHT AGAINST... SOMETIMES IT'S SO ABSURD. KEEPING SOME DEGREE OF LEVITY ISN'T JUST THE BEST WAY TO DEAL, IT SHOWS OTHERS HOW *SILLY* IT IS.
I MEAN, SCIENTISTS ARE TELLING US WE'RE ALL GONNA DIE IF WE DON'T ACT.

WHY ARE WE TALKING ABOUT ANYTHING *ELSE?* IT'S *RIDICULOUS.*
FACING EXTINCTION'S A BIG DEAL, YOU KNOW?

Someone who's helped get Antarctica off-limits for mineral exploration for 50 years, who's trying to install protections in international waters--he still believes...
Having a sense of **humor** can change the world. Telling the right **story** can change the world.
So just imagine what's possible...

If we all decided to do just a **little** more?
DEVELOPED WITH
DIMA LITVANOV
Dima's Story
WRITTEN BY
PORNSAK PICHETSHOTE
ART BY
PETER GROSS
COLORS BY
JP BOVE
LETTERING BY
BERNARDO BRICE

Meat Free Monday

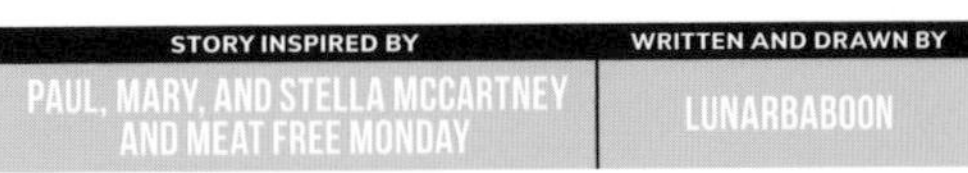

STORY INSPIRED BY
PAUL, MARY, AND STELLA MCCARTNEY AND MEAT FREE MONDAY
WRITTEN AND DRAWN BY
LUNARBABOON

"NOTHING GROWS HERE, THEY TOLD US.
"BUT WE DREAM.
"WE HOPE.
DEVELOPED WITH
GEORGE MONBIOT
WRITTEN BY
ROB WILLIAMS
ART & COLORS BY
BEN OLIVER
LETTERING BY
BERNARDO BRICE
OWN THE SKIES
"AND WE *WORK*."

RAIN PLEASE

SKREE
ATTENTION!
BE AWARE! THE CLOUD WHOSE IRRIGATION YOU ARE CURRENTLY PROFITING FROM IS OWNED BY THE FOODSTUFFED™ CORPORATION OF CALIFORNIA, UNITED STATES.
ALL DANCING WITHIN THE RAIN SHALL CEASE FORTHWITH UNDER THE STATUTES OF CALIFORNIA LAW!
WHAT? YOU CAN'T OWN A CLOUD...
FOODSTUFFED™ CAN AND *DO*, MADAM. FOUR MONTHS AGO WE BOUGHT ALL AERIAL LAND OWNERSHIP OF THIS ENTIRE STATE. I REFER YOU TO SUBSECTION 42, CLAUSE (13 XVIII).
WHAT?
ANY CROPS YOU CULTIVATE HERE HAVE BEEN GROWN WITH WATER SUPPLIED BY FOODSTUFFED™ CLOUDS AND THEREFORE, THEY ARE OWNED BY FOODSTUFFED™.
BUT...
THAT'S CRAZY...
"The first man who, having fenced in a piece of land, said 'This is mine,' and found people naïve enough to believe him, that man was the true founder of civil society. From how many crimes, wars, and murders, from how many horrors and misfortunes might not any one have saved mankind, by pulling up the stakes, or filling up the ditch, and crying to his fellows: Beware of listening to this impostor; you are undone if you once forget that the fruits of the earth belong to us all, and the earth itself to nobody."
- Jean-Jacques Rousseau

CONSUMED TO DEATH

BY GEORGE MONBIOT
ILLUSTRATION BY IAN STOPFORTH

Consumerism, which we're being told enhances our lives, makes us richer and makes our lives better, actually destroys our peace of mind, self-worth and sense of security.

It constantly creates new needs, new wants, which didn't exist before.

And once we want that thing, our lives feel incomplete until we've obtained it. We forget about what the impact might be on the people of future generations, or on people who live on the other side of the globe.

When we start to monetize our relationships, even with the natural world, everything becomes consumable.

It's not enough to exist as a consumer alone. We desperately need to reconnect with other people and the physical world, the natural world once more. Now is a great time to reassess who we are and where we stand.

To re-evaluate our relationships with each other and the rest of the living world.

To rekindle our moral imaginations, to reconnect and to overthrow consumerism!

Help by pausing for a second:

IS THE OBJECT YOU ARE ABOUT TO BUY SOMETHING THAT YOU REALLY NEED, OR JUST WANT?

HOW MUCH DID THAT COST: MAKE-UP

STORY INSPIRED BY	WRITTEN BY	ART & LETTERING BY
CHEDDAR GORGEOUS	DAVE SCHNEIDER & AMBER WEEDON	WAR AND PEAS

War and Peas

HOW MUCH DID THAT COST: PHONES
OooOOOOoh... LOOK AT YOUR PHONE. HOW MUCH DID THAT COST?
DEVELOPED WITH
PROF. BEN GARROD
WRITTEN AND DRAWN BY
DREW SHENEMAN

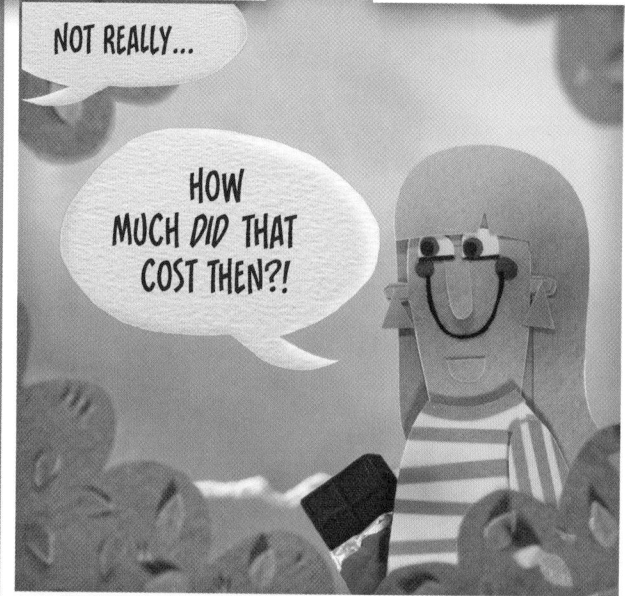

DEVELOPED WITH
PROF. BEN GARROD
WRITTEN BY
PAUL GOODENOUGH
ART BY
DANIELLA ORSINI

PRAWN FREE
THE PRAWN FAMILY ARE GOING ON HOLIDAY! BUT ARE THEY SKATE-ING ON THIN ICE?
FUNNY FISH FARM
PASSPORT
DEVELOPED WITH ANTON LITVANOV
WRITTEN BY JAMIE WOOLLEY
ART BY DAVEY JONES
LETTERING BY BERNARDO BRICE
CHICKEN OR FISH?
JAWS
I'D KRILL FOR SOME FISH!
COR! OCEANS ARE PLUNDERED TO CATCH FISH AND KRILL, JUST TO FEED FARMED PRAWNS AND FISH!
Holiday Finn
HOORAY! WE FINALLY MADE IT!
BLIMEY, IT'S LIKE A TIN OF SARDINES!
POOR JUNIOR! BUT PRAWN AND FISH FARMS ARE OFTEN OVERCROWDED!
YOU'VE GOT TO BE SQUIDDING! THIS IS DISGUSTING!
PARP
CHUFF
YUCK! POLLUTION FROM AQUACULTURE FARMS IS DANGEROUS!
WHAT A CLAM-ITY! WE'RE ALL SICK FROM THAT FILTHY POOL! I'LL CALL FOR THE STURGEON!
BARF! DISEASE SPREADS LIKE BILLY-O, INFECTING WILD CREATURES AS WELL...
I WON'T CARP ON, BUT YOU SHOULD BE MORE CAREFUL!
...AND OVERUSED ANTIBIOTICS CAN CREATE LETHAL SUPERBUGS!
MAYBE A STROLL WILL PERCH US UP?
THIS IS A STICK-UP! GIVE ME ALL YOUR SHELLS!
FOR THE LOVE OF COD, NOT MY SHELL!
FISH AND SEAFOOD ARE SHIPPED AROUND THE WORLD FOR SHELLING AND PROCESSING, AND SHIPPED BACK AGAIN TO BE SOLD! THAT'S JUST C-RAY-ZY!
WE SHOULD HAVE STAYED AT HOME!
CRIKEY READERS, SOUNDS LIKE A SLAP-UP FISH SUPPER ISN'T ALWAYS SUCH A BRILL-IANT IDEA!
BUT YOU CAN BECOME A DAB HAND AT CHOOSING SUSTAINABLY CAUGHT AND PRODUCED SEAFOOD INSTEAD!

Aware Animals

Away

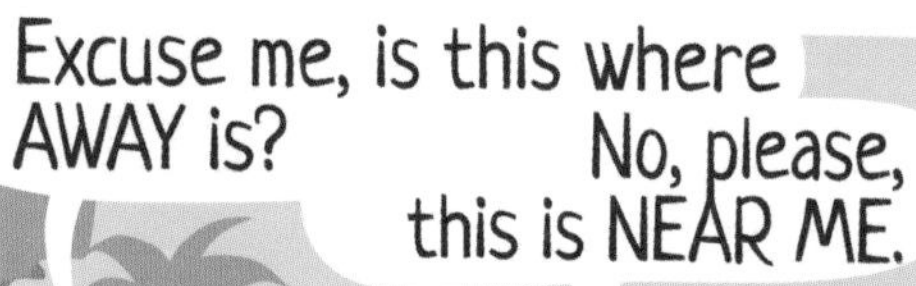

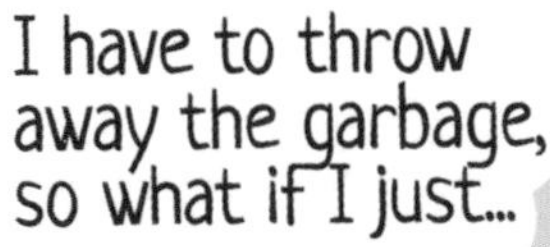

WRITTEN AND DRAWN BY
AWARE ANIMALS

PURCHASING POWER
DEVELOPED WITH
MICHELLE DESILETS & HELEN BUCKLAND
WRITTEN BY
PAUL GOODENOUGH
ART & COLORS BY
HECTOR TRUNNEC

Eat healthy Eat wise

LET'S MAKE THIS LAW

Energy 513kJ 121kcal	Fat 1.4g	Saturates 0.3g	Sugars 8.4g	Salt 0.2g	Enviro Impact 0/5
6%	2%	2%	9%	3%	Unsustainable Palm Oil 0.2g

WRITTEN BY
PAUL GOODENOUGH
POSTER DESIGNED BY
MOKSHA CARAMBIAH

IS THE PRICE RIGHT?
DEVELOPED WITH
WILL TRAVERS OBE
WRITTEN BY
JAMIE WOOLLEY
ART & LETTERING BY
STILLY
I'D LIKE TO BUY YOUR LAND TO BUILD A CAR PARK.
REALLY? HOW MUCH ARE YOU GOING TO GIVE ME FOR IT?
I WAS THINKING $100.
HMM. I'M NOT SURE. HOW DID YOU WORK THAT OUT?
THAT'S WHAT I RECKON IT'S WORTH. SUPPLY AND DEMAND, RIGHT?
I SEE. DOES THAT INCLUDE THE FLOWERS?
FLOWERS? OF COURSE NOT. I DON'T NEED FLOWERS. WHY WOULD I PAY FOR THEM?
BUT BEES NEED THEM. THEY SEEM TO VALUE THEM. AND WHAT ABOUT THE TREES?
TREES AREN'T ESSENTIAL FOR CAR PARKS!
TRUE. THEY'RE PRETTY VALUABLE THOUGH, PRODUCING OXYGEN, SUPPORTING INSECTS AND ANIMALS STABILIZING THE SOIL...
I JUST WANT THE LAND.
YOU MEAN THE SOIL? FULL OF ORGANIC MATTER, STORING NUTRIENTS AND CARBON, ENABLING THE TREES AND FLOWERS TO GROW AND FEED US ALL.
I DON'T THINK YOU'RE GETTING THIS.
SO IF WE ADD ALL THAT UP, AS WELL AS THE WATER PURIFICATION, NUTRIENT CYCLING, AND THE HABITATS FOR ENDANGERED SPECIES, NOT TO MENTION THE FACT MY FAMILY HAS LOOKED AFTER THIS LAND
HUNDREDS OF YEARS, SO IT HAS...
I CAN'T COME UP WITH A FIGURE FOR THAT!
NOW YOU SEE WHERE THIS IS GOING.
STILLY

HOW MUCH IS LAND *REALLY* WORTH?

Inspired by Will Travers OBE and Diane Coyle. Written by Paul Goodenough.

You see it across the world—undeveloped land going cheap. Undeveloped, or wild, land is sold at a fraction of the price of land that's already been developed on. What this does is drive businesses and individuals to opt to destroy wild land because they can turn a far greater profit than redeveloping existing, expensive, developed land.

And when they do, the cost to the planet, and us, is devastating.

WHY DOES THIS HAPPEN?

Currently, land purchase price is based upon factors like ease of access, potential for commercialization, restrictions, amenities etc—essentially, how much likely profit can humanity get from the land if they were to transform it into human use, and away from its intrinsic value and ecosystem services it provides in its natural state.

There is almost no consideration to the value the land is already giving. Even to the outputs that directly benefit humanity such as air quality, aiding pollination, carbon sequestration, flood risk reduction, and improving our physical and mental health.

When you add in factors like biodiversity, providing homes to wildlife, maintaining wildlife corridors and habitat interconnectivity, we soon realize the true value of undeveloped and wild land.

We need to change the way land is valued. We need to ensure the planetary value is always factored into every land purchase and the loss to the planet is considered every time land is developed.

If we do this, businesses across the planet would pay more to develop on wild land than they would to redevelop existing land, making regeneration on developed land a more attractive option.

This would not only stop the wholesale destruction of our wild places, but it would also incentivise the re-development of millions of deserted areas like old factories, theme parks and derelict houses – improving our villages, towns, and cities as well.

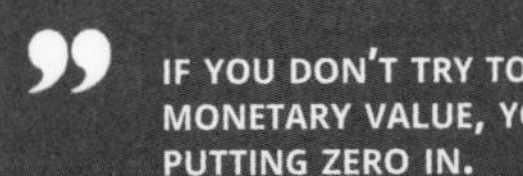

> **IF YOU DON'T TRY TO ADD A MONETARY VALUE, YOU'RE PUTTING ZERO IN.**
>
> **AND THAT IS DEFINITELY THE WRONG ANSWER.**
>
> **Diane Coyle** is a professor at the University of Cambridge, leading a project to ascertain the capital value of nature.

Join us in calling for this to become law:

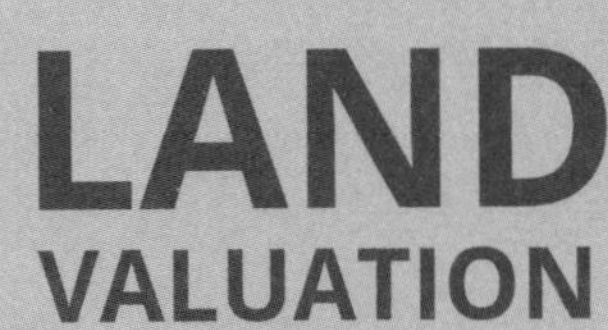

DYE ANOTHER DAY

WRITTEN BY LUCY SIEGLE & PAUL GOODENOUGH
ART & LETTERING BY GOODBADCOMICS

3423 CALLS LATER...

NOT ALL JEANS GO TO HEAVEN

DEVELOPED WITH	WRITTEN BY	ART BY
CANDIANI	SARAH FLORENCE LORD & PAUL GOODENOUGH	THINGS IN SQUARES

3,000 LITERS OF WATER WENT INTO MAKING THESE POLLUTING, HUMAN- AND ANIMAL-POISONING, CARBON RELEASING, MICROPLASTIC-CONTAMINATING STRETCH JEANS, BUT...

I LOOK GORGEOUS!
TIL DEATH DO US PART.

LATER
ONLY THINGS THAT DIE GET INTO HEAVEN. BUMMER, PANTLESS FOR ETERNITY.
SHOULD'VE WORN COMPOSTABLE JEANS

MUCH LATER
PLEASE KILL ME

THINGSINSQUARES

Cruelty What

WRITTEN AND DRAWN BY
WAR AND PEAS

War and Peas

Fabulous

WRITTEN AND DRAWN BY
WAR AND PEAS

War and Peas

HOW MUCH OF THE EARTH'S SURFACE IS FORESTED
Humanity has been the planet's foremost inventor, hunter, and farmer since we discovered fire, one of our early triumphs we're most proud of.
Of course, being the apex hunter comes with certain responsibilities, which we take seriously.
Consider: mankind is the only species ever to actively save or protect another species.
Some of our most recent highlights include:
THE PEREGINE FALCON!
With its haughty bearing and beautiful cruel eyes, a top speed of 200 mph makes this king of the skies the fastest animal in the world.
Until a human steps into a fighter jet!
THE RHINO!
Although they might look ugly and prehistoric, Rhinos are one of Africa's "big five" and therefore a valuable draw for high-spending tourists.
So rest easy, horny, we've got your back!
THE WHALE!
Whether they're toothed or baleen, you can't help but be moved by whales' poetic loveliness.
And guess who's doing their damnedest to save them!
Humanity: Annual Report
Yes, we're proud of protecting our fellow tenants of Mother Earth, from the cutest panda to the fiercest cougar.*

MICHAEL,
SEE THIS LEAKED REPORT
"HUMANITY INC." IS PLANNING
TO PUBLISH BY THE END
OF THE QUARTER. THIS IS
EVERYTHING THAT'S WRONG
WITH CORPORATE CAPITALISM.
THE TIGER!
Not for nothing, does this sexy hunter adorn posters, flags and murals. It's a real looker.
Tyger, tiger burning bright indeed!
THE SEA OTTER!
Unlike their often slimy cousins the river otter, these sweethearts are so cuddly they wouldn't look out of place snuggling next to your kid at night.
Even though in truth they'd probably bite half of your little kid's face off, we still think they're worth saving!
EXCREMENTUM ANTHROPHAGOS!
*Of course to save these awe-inspiring creatures we've had to downsize a little. Some unproductive elements have been let go. Creatures who would never grace a poster or become a corporate logo! We've saved the whale and the white tiger, do you really care if few blind insects who live their lives in other creatures' s@*t survive?
Marketability Rating: A. B. C. D.
PLEASE WORK WITH
SHARON AT COMMS ON
HOW TO BEST COUNTERACT IT
AND GET BACK TO ME.
WRITTEN BY
PETER MILLIGAN
ART & LETTERING BY
GUILLERMO ORTEGO

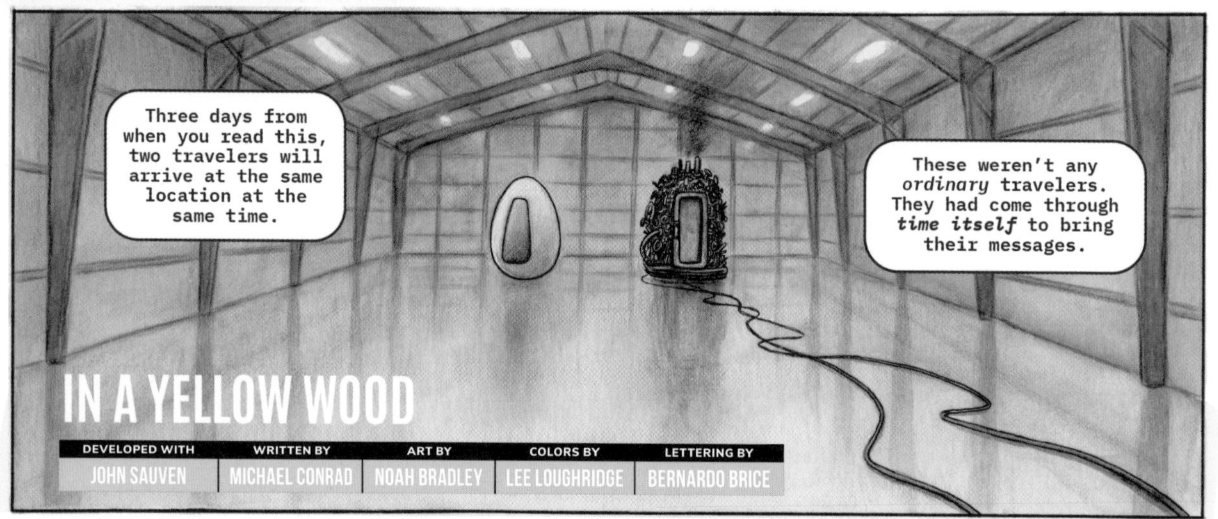
Three days from when you read this, two travelers will arrive at the same location at the same time.
These weren't any *ordinary* travelers. They had come through ***time itself*** to bring their messages.
IN A YELLOW WOOD
DEVELOPED WITH JOHN SAUVEN
WRITTEN BY MICHAEL CONRAD
ART BY NOAH BRADLEY
COLORS BY LEE LOUGHRIDGE
LETTERING BY BERNARDO BRICE

It was unclear at first why they had made the journey. Initial responses were fearful, but they soon made their reasons for visiting clear.
VISSSS
SHUNK
WOOOSSH

Their time was limited; each was eager to share the information they had risked their lives to bring to us.

As we scrambled to set up recording devices, one of the travelers began to speak.

This was ***not*** good news at all.
PSSSSHH

≈COFF≈
THE FUTURE IS BLEAK.
WE AS A PEOPLE HAVE FAILED THE PLANET.

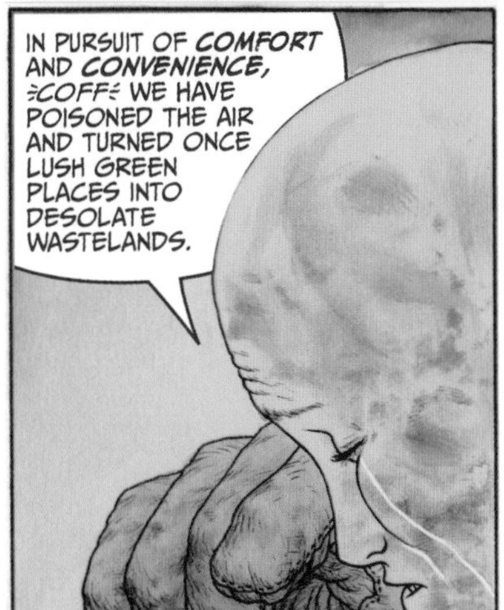
IN PURSUIT OF COMFORT AND CONVENIENCE, ≈COFF≈ WE HAVE POISONED THE AIR AND TURNED ONCE LUSH GREEN PLACES INTO DESOLATE WASTELANDS.

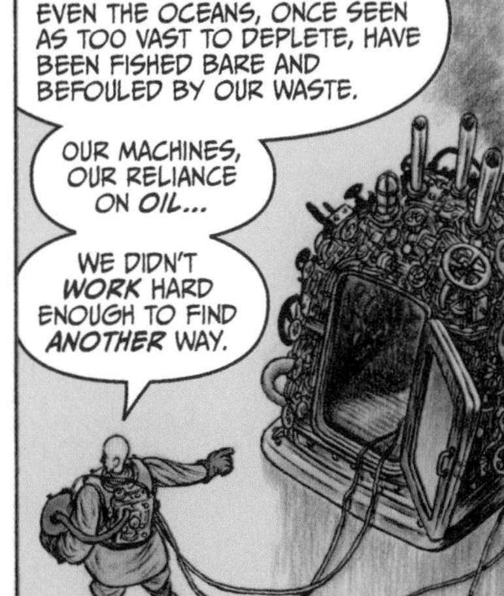
EVEN THE OCEANS, ONCE SEEN AS TOO VAST TO DEPLETE, HAVE BEEN FISHED BARE AND BEFOULED BY OUR WASTE.
OUR MACHINES, OUR RELIANCE ON OIL...
WE DIDN'T WORK HARD ENOUGH TO FIND ANOTHER WAY.

I HAVE RETURNED TO THIS CRITICAL TIME TO BEG YOU--

--PLEASE, MAKE THE FOLLOWING A PRIORITY.
THE GREATEST MINDS OF OUR TIME HAVE FOUND POTENTIAL SALVATION IN TWO SIMPLE STEPS.
THESE WERE KNOWN, EVEN BEFORE YOUR TIME, BUT WERE NOT ACTED ON.
YOUR FAILURE WILL BECOME YOUR CHILDREN'S BURDEN.

THE FIRST IS SIMPLE: A FOCUS ON FINDING CLEAN AND RENEWABLE SOURCES OF ENERGY.
YOU HAVE THE TECHNOLOGY: A DIRECT FOCUS MUST BE GIVEN TO MAKING THESE THE NORM IF WE WISH TO SURVIVE.
THE OTHER IS TO MOVE TOWARD A MORE PLANT-BASED DIET. ⸗COFF⸗
THIS MAY SOUND HARD, BUT REALLY, IT'S CRUCIAL. INDUSTRIAL MEAT PRODUCTION WAS MUCH MORE HARMFUL THAN WE COULD HAVE IMAGINED.

THE LAND REQUIRED ⸗COFF⸗ THE ENERGY EXPENDED ⸗COFF COFF⸗ WE WEREN'T SEEING THE BIG PICTURE-- THE SUFFERING...

I MUST SIT.
⸗COFF⸗ GOD, THE AIR HERE IS INCREDIBLE...

I TOO COME FROM THE FUTURE.
I TOO BRING NEWS OF THE WORLD TO COME.

I COME TO SAY MUCH THE SAME. THIS OTHER TRAVELER IS CORRECT.
THESE TWO STEPS ARE VASTLY IMPORTANT. THERE ARE OTHERS, BUT THESE TWO ARE ESSENTIAL, THESE ARE THE FUNDAMENTAL CORE OF POSITIVE CHANGE.
CLIK!
CLIK!

I BRING HOPE.
HUMANITY WILL DO WHAT'S RIGHT. A RENEWAL OF GREEN PLACES CAN AND WILL OCCUR. A MORE HARMONIOUS EXISTENCE WILL HAPPEN.
RESTORATION... OF LAND AND SEA.

WHEN?! COFF WHEN WILL THIS HAPPEN? WHAT TIME DO YOU COME FROM?!

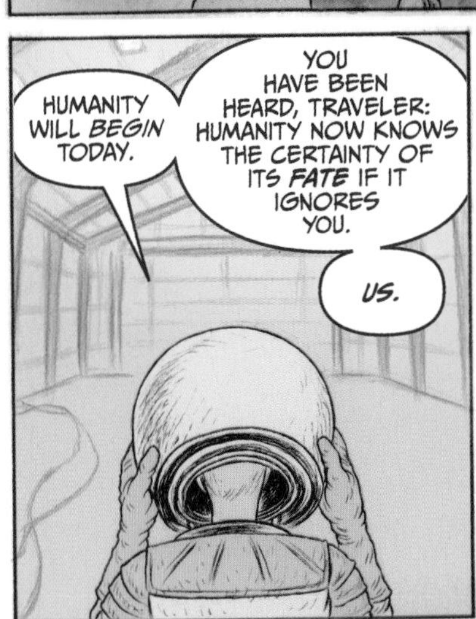
HUMANITY WILL BEGIN TODAY.
YOU HAVE BEEN HEARD, TRAVELER: HUMANITY NOW KNOWS THE CERTAINTY OF ITS FATE IF IT IGNORES YOU.
US.

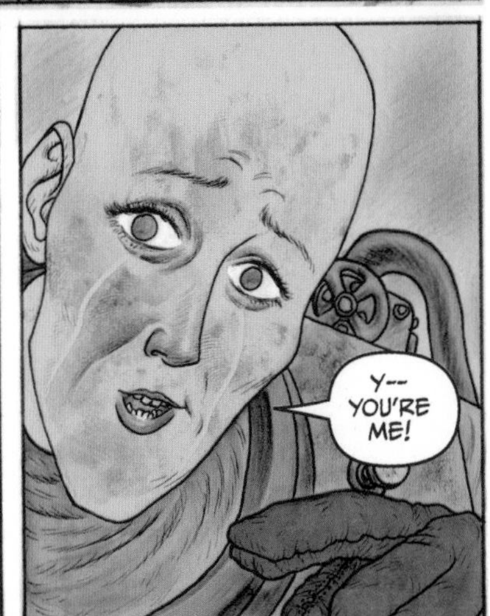
Y-- YOU'RE ME!

YES. THE WORLD YOU SHALL RETURN TO WILL HAVE BEEN CHANGED. YOU WILL HAVE BEEN CHANGED.

"BUT THERE IS MUCH LEFT TO BE DONE. WE CAN COUNT ON THE PEOPLE IN THIS TIME TO DO WHAT'S RIGHT."
They didn't have much time left, and neither would we if we didn't heed their warning.
Three days from when you read this, two travelers will arrive at the same location at the same time.
We will have already begun to change.

"Chen *Tai,* I don't normally take a meeting like this but my cousin says you're practically family and it's *urgent.*"
"Thank you, Ah Wong. My other supplier fell through and I have some wealthy clients in Shenzhen I can't disappoint."
有限公司
PALACE JEWELLERY LTD
冠超珠寶行
JUST BUSINESS
STORY INSPIRED BY
STEVE BACKSHALL
WRITTEN BY
AMY CHU
ART & COLORS BY
E.J. SU
LETTERING BY
BERNARDO BRICE

Ah, cops are clamping down. Rhino horn, tiger bone, pangolin scale...
What's next? It's getting so you can't even enjoy a bowl of shark's fin soup anymore.

Traditional Chinese medicine has been around for 5,000 years. Westerners say there's no proof it works. Who needs proof when you have thousands of years of fine tuning?
Belief is a powerful thing. I have a client who has cancer. Who knows? People who have money will pay anything. Personally, if I'm that sick I go to a Western hospital. But Chinese medicine. Just in case.

My wife is superstitious. Me, I'm practical.
S'why I'm Buddhist *and* Christian. Just in case.

So about that rhino horn...?
Listen, you seem like a nice lady. Rhino horn is very, very expensive. And *very* illegal. The going rate is over $60k per kilo.

!
So much?
Supply and demand. It's just *business.* When there was still rhino and tiger in *China,* they were still expensive. It was never for the peasants.
You have *cash.* I have lots of other things. Come inside. I give you a good deal.

What's he messing with that snake for? @#$%$ crazy Westerners. Guys like that are bad for business.
Hypocrites. They use feng shui and acupuncture but don't believe in Chinese medicine?
These gwei lo think chicken soup cures a cold.
Yet everyone knows it's the chicken *and* the ginseng.

Do you have any tiger bone?
Heh, right here. I'm a tiger! *At least,* that's what my wife tells me.

What about this? *Real?*
Of course not. You think I'd have a real rhino cup out like that?

Ming Dynasty. Our ancestors killed them all off to make trophies like these.
Just like the Americans and their bison, the Europeans and their dodo birds.

Look, I'm not here for souvenirs or a history lesson. You don't have what I need. I will find someone else.
Okay, okay, here, here let me show you something.

Beautiful, yeah? Just in from South Africa. All front horn.
How many do you have?

I have eight. All over seven kilos each.
Eight?! How did you get so many?

Connections.
At the airport. Overseas. Government officials. I'm an *important* man, Mrs. Chen.

How do I know they're not fake?
If I sell you a fake, my reputation is ruined. Here, I'll show you...

You won't find a better deal all around, Chen Tai. I'm an honest man.
So. How many would you like?

AIR POLLUTION CAUSES LUNG CANCER. THAT'S A FACT.

WRITTEN BY
PAUL GOODENOUGH
ART BY
PATRICK GODDARD
COLORS BY
DYLAN TEAGUE
LETTERING BY
JIM CAMPBELL

BUT FOR YEARS, THE INDUSTRY DENIED IT...

...CHALLENGED EVIDENCE WITH PAYROLLED EXPERTS. SEEDED DOUBT. ARGUED. DELAYED...ANYTHING THEY COULD...

PARIS

PARIS

...TO CONTINUE BUSINESS AS USUAL. TO KEEP THE PROFITS ROLLING IN...

...WHILE THEY **KNEW** THEY WERE KILLING US.

BOY

HAPPY BIRTHDAY

IS FAST FASHION MAKING US SICK?

Every time we wash 5 kilograms of synthetic clothes, 9 million micro- and nanofibers are released into our streams and oceans and end up in the fish we eat. The fibers from the clothes we wear are released into the very air in our homes. We breathe them in, all day…every day.

Scientific research links nylon and polyester fibers with reduced lung cell growth. Just how much damage they're doing is only now being uncovered. And the industry is terrified the truth will come out…

And that truth is: the fast fashion industry has a range of solutions. But they would rather make more profits than use them — and that's why you, me, and the rest of the world are inhaling their fibers.

Demand fast fashion stops trading our health for their profits.

Fashion should be safe for us all.
#sickfashion #makefashionsafe

DEVELOPED WITH
THE PLASTIC SOUP FOUNDATION
POSTER DESIGNED BY
MOKSHA CARAMBIAH

I'M GUY BYRNE-WOODS, CEO OF THE FORESTRY FELLING SYNDICATE.
TIK... TOK...
GUY BYRNE-WOODS FFS

ON BEHALF OF THE GLOBAL LOGGING INDUSTRY, I WANT TO THANK PEOPLE LIKE YOU FOR COLLECTIVELY INVESTING HUNDREDS OF MILLIONS WITH US.
TIK... TOK...

OUR PENSION FUNDS, NO MATTER HOW PIDDLY, ALL ADD UP.
TIK... TOK...

AND WITH ALL YOUR CASH...
TIK... TOK...

TIK... TOK...
... WE'VE BEEN ABLE TO DESTROY MORE NATURAL HABITAT THAN WE EVER THOUGHT POSSIBLE.

FROM ALASKA TO THE AMAZON...
TIK... TOK...

...FROM SIBERIA TO SUDAN.
TIK... TOK...

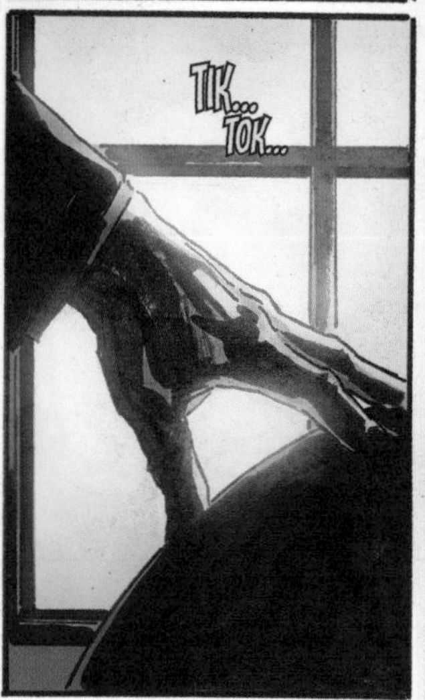
TIK... TOK...

TIK... TOK...
SYNERGY.

TIK...
TOK...
WE'VE ALSO USED YOUR INVESTMENTS TO FIGHT PESKY GOVERNMENT RESTRICTIONS.

IN FACT, EVERY SINGLE MINUTE, WE MANAGE TO CLEAR ABOUT 30 FOOTBALL FIELDS OF FOREST.
TIK...
TOK...

TIK...
TOK...

TIK...
BONG

THERE GOES ANOTHER ONE!
TIK...
TOK...
BUT WE CAN ONLY KEEP UP OUR RECORD-BREAKING WORK WITH YOUR HELP.

WITH YOUR CONTINUED INVESTMENT WE CAN DO EVEN MORE TOGETHER NEXT YEAR--
TIK...
TOK...
TO MAKE MONEY, AND DESTROY THE PLANET.

TIK...
TOK...

SO REMEMBER, YOUR MONEY, IS LITERALLY CHANGING THE WORLD.

DO YOU KNOW WHAT YOUR MONEY IS FUNDING?

WRITTEN BY LUCKY GENERALS
ART BY DOUG BRAITHWAITE
COLORS BY SEBASTIAN CHENG
LETTERING BY BERNARDO BRICE

PENSION POWER!

BY MAKE MY MONEY MATTER

The choice on where to keep, save, and invest your money is more powerful than you think.

In fact, deciding the pension we hold is one of the most powerful things we can do to cut our carbon, tackle the climate crisis, and give our money a chance to grow in the long run.

HOW'S THAT?

In the UK alone, there's **£3 trillion** held in pensions, and it's money that many of us have the ability to control.

But while investing in many vital businesses, pension funds have also been funding some of the nastiest stuff on the planet.

From tobacco to fossil fuels, weaponry to deforestation, pension funds have invested trillions on our behalf without ever asking us the crucial question:

Do these investments create a world we actually want to live in?

This means our money may be contradicting our lifestyles, with climate campaigners invested in coal, vegans invested in factory farming, and scientists invested in tobacco.

MAKE YOUR MONEY MATTER!

But it doesn't have to be this way.

You can make your money matter, and ensure the trillions of pounds invested on our behalf builds a better world.

After all, what's the point in saving for retirement in a world on fire?

Find out more at www.makemymoneymatter.co.uk

MakeMyMoneyMatter.co.uk

WHILE YOU SLEEP

DEVELOPED WITH
RICHARD CURTIS AND VICTOR SOLÍS
WRITTEN BY
PAUL GOODENOUGH
ART BY
ABHIJEET KINI

DEAR, HE DOES HAVE A POINT.

Woke

War and Peas

DEVELOPED WITH RICHARD CURTIS

WRITTEN AND DRAWN BY WAR AND PEAS

In my town, there was a hole...
Wish there was another job, but what can you do...
WRITTEN BY
CHEDDAR GORGEOUS, PAUL GOODENOUGH, JOHN MCNAMEE
ART & LETTERING BY
JOHN MCNAMEE

Yeesh!
That doesn't look good Dad.
CRACK!
CRACK!

Need more money...

Let's get out of here!
CRSH!
CRACK!

Ahh... Much better.
Those extra shifts really paid off.

Daaad...
"CRACK!"

We'll just move further!

But...
The Hole: are things really that bad?
HERE Too?!
"CRACK!"

Come on!
CRACK
CRK!

Are you kidding me?!
Grab a shovel son. We need the BIG BUCKS!
CRACK!"

Ahh...
It's amazing Dad! I can't even see the Hole.
TOWN LIMIT

We're finally safe...
"CRACK!"

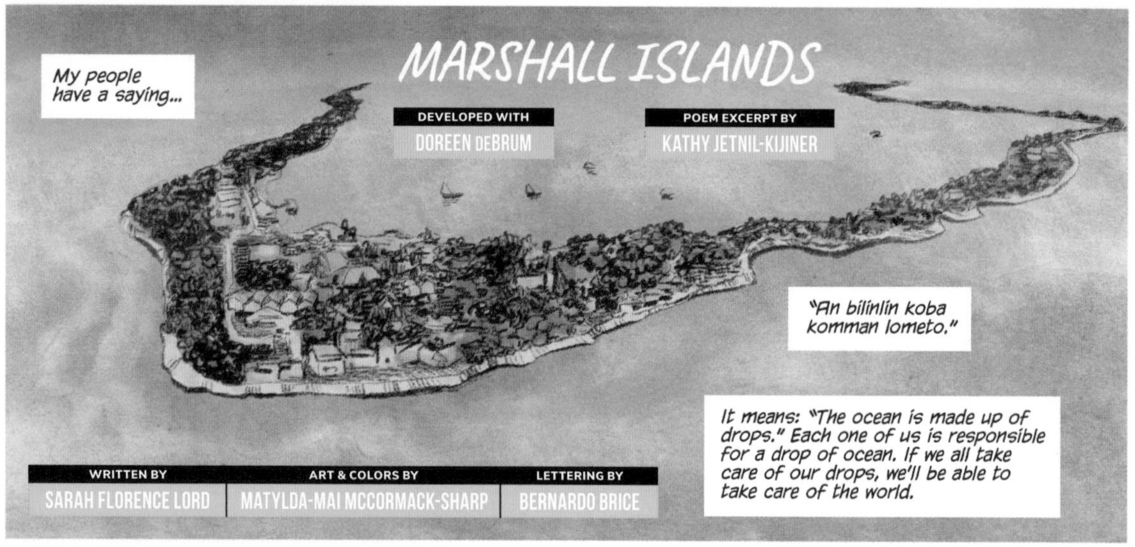
My people have a saying...
MARSHALL ISLANDS
DEVELOPED WITH
DOREEN DEBRUM
POEM EXCERPT BY
KATHY JETNIL-KIJINER
"An bilinlin koba komman lometo."
It means: "The ocean is made up of drops." Each one of us is responsible for a drop of ocean. If we all take care of our drops, we'll be able to take care of the world.
WRITTEN BY
SARAH FLORENCE LORD
ART & COLORS BY
MATYLDA-MAI MCCORMACK-SHARP
LETTERING BY
BERNARDO BRICE

In 2019, I left my homeland of the Marshall Islands to become the UN ambassador in Geneva, to bring the issues faced by my people to the global stage.
If you don't know already, the Marshall Islands is in immediate threat from rising sea levels. If we don't do something, our islands... My home... Will be underwater, and we will all be homeless.

The committee decided it would be me who should go to Geneva and set up the mission.
It was only meant to take six months...

I barely had time to say goodbye to my family...

I was suddenly starting a new life in a foreign country... 14,000 km from home with no money, no transportation and no accommodation.
It was a... difficult... first few months.

When I arrived I immediately began campaigning for us to become a member of the Human Rights Council.
I really struggled to see how I could make people understand how serious the threat to our islands is, how I could make them relate to the issues that my people face.

But I was determined.

All of our islands will be underwater within 60 years. We've already lost one, we simply cannot wait any longer to take action...
ILES MARSHALL

If my people become climate refugees what does that mean? Where do we go? As a nation, what do we **become?**

I left my homeland so that my people would not be forced to.

Without our islands we are nothing.

DOREEN DE BRUM
So here I am in Switzerland, representing my homeland, a tiny group of islands in the Pacific Ocean, on the global stage.
DOREEN DE BRUM
When we launched our embassy the offices were not ready yet, we had nowhere to hold the event...
So we decided to hold it on the shore of Lake Leman.

And we had a Korkor, a traditional Marshallese canoe, brought from the Pacific Ocean to the Swiss Alps. It was a beautiful, poignant reminder that no matter where we live, climate change affects us all.
And I was reminded of the expression, "with climate change, we are not all in the same boat, but sooner or later we all must weather the same storm."

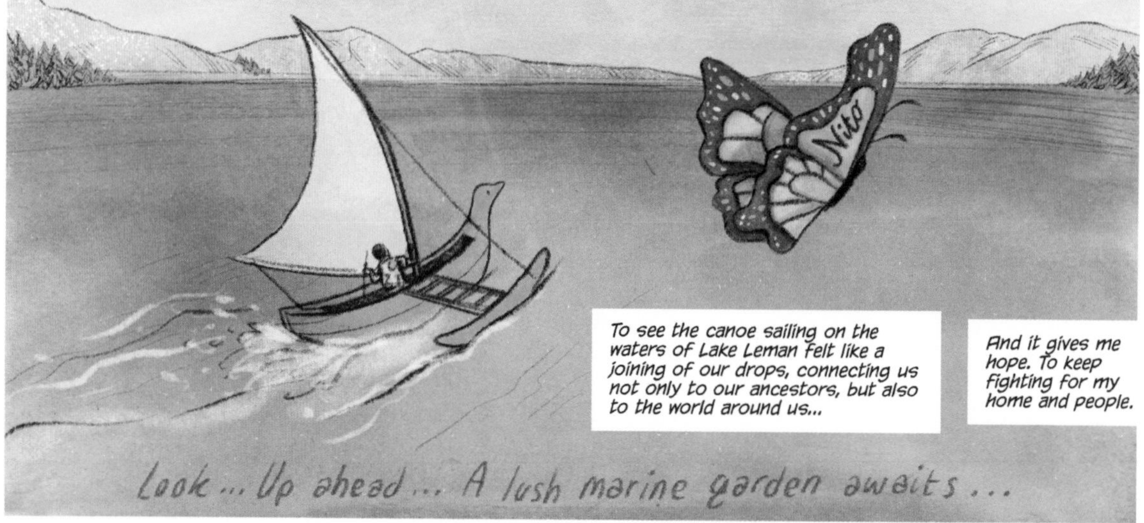
Nito
To see the canoe sailing on the waters of Lake Leman felt like a joining of our drops, connecting us not only to our ancestors, but also to the world around us...
And it gives me hope. To keep fighting for my home and people.
Look... Up ahead... A lush marine garden awaits...

TED TALKS

STORY INSPIRED BY	WRITTEN BY	ART & LETTERING BY	WITH THANKS TO
PETER GABRIEL AND THE INTERSPECIES INTERNET	PAUL GOODENOUGH	SARAH GRALEY	TED

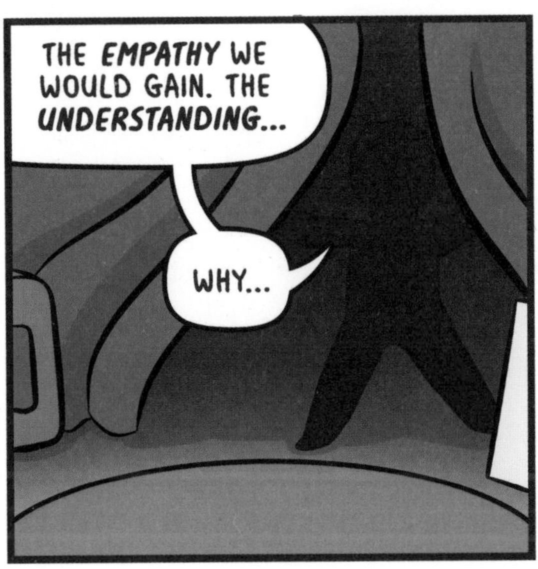

"I'VE BEEN WALKING FOR ABOUT 30 YEARS WITH OUR INDIGENOUS ELDERS--COMING TO INSTRUCT US, HELPING US REMEMBER HOW TO WALK IN THE ORIGINAL WAY OF LIFE AND TO FOLLOW THE PROTOCOLS OF THE EARTH HERSELF.

"IT IS TIME FOR US TO GALVANIZE AND COLLABORATE WITH INDUSTRIES, PHILANTHROPISTS, GOVERNMENTS, AND ORIGINAL PEOPLE TO FOLLOW THESE INSTRUCTIONS.

"WHAT WE WANT IS TO BRING BACK CONSCIOUSNESS. EVERY CULTURE HAS AN ACTIVE OBLIGATION TO COME TOGETHER TO SOLVE THIS CRISIS.

"TO BRING ABOUT CHANGE IN OUR ENVIRONMENT AND OUR SOCIETY. REMEMBER THAT IT IS THE COMMUNITIES CLOSEST TO MOTHER EARTH AND NATURE THAT ARE THE ONES MOST AFFECTED.

"A TABLE HAS BEEN SET, AND WE'RE ALL BRINGING THESE THREADS THAT MOTHER EARTH HERSELF HAS CULTIVATED.

"THERE ARE UNBROKEN THREADS THAT HAVE BEEN BROUGHT FROM THE ORIGINAL DAYS THROUGH OUR INDIGENOUS RELATIVES THAT WILL GUIDE THE WEAVING PROCESS, SO WE CAN BEGIN TO SEE THE NEW TAPESTRY OF WHAT WE'VE COME TO SERVE.

DEVELOPED WITH
JYOTI MA AND MINDAHI BASTIDA
ADAPTED BY
SARAH FLORENCE LORD & TYLER JENNES
ART & COLORS BY
ZOE THOROGOOD
LETTERING BY
BERNARDO BRICE

"THERE ARE MANY PROPHECIES THAT HAVE BEEN PASSED THROUGH ORIGINAL PEOPLE'S LINES AND THEY HAVE PREPARED US FOR THE TIME WE'RE STANDING IN NOW.

"THOSE PROPHECIES ARE NOW INSTRUCTING US HOW TO WALK THROUGH THIS. MOTHER EARTH IS LITERALLY SPEAKING TO US IF WE JUST LISTEN.

"ACCORDING TO OUR LORE, THIS IS THE EARTH'S SIXTH EXTINCTION. SHE KNOWS HOW TO HANDLE IT IF WE JUST LET HER.

"SHE HAS SENT HER ORIGINAL PEOPLE TO HELP US THROUGH THESE TIMES, AND WE MUST RADICALLY LISTEN TO HER INSTRUCTION.

"WHEN YOU SEE A RAINBOW, YOU SEE THE COLORS. IF ONE COLOR IS MISSING, IT IS NO LONGER A RAINBOW."

LESSONS FROM THE LAKOTA
DEVELOPED WITH
MOSES BRINGS PLENTY
WRITTEN BY
SARAH FLORENCE LORD
ART & COLORS BY
ZOE THOROGOOD
LETTERING BY
BERNARDO BRICE
THERE IS A SAYING IN OUR CULTURE...

"LÉ WANJI PHEŽÍ OTHÓKĄHE MAKHÁ WIČHÓNI."

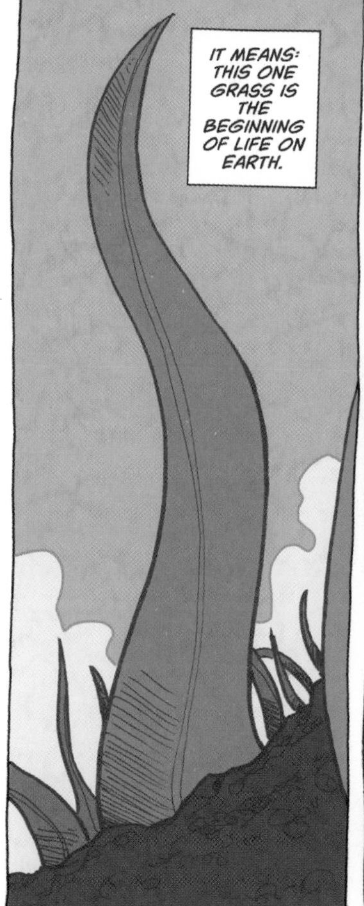
IT MEANS: THIS ONE GRASS IS THE BEGINNING OF LIFE ON EARTH.

IF THE GRASS IS HEALTHY THEN THE BISON ARE HEALTHY. IF THE BISON ARE HEALTHY THEN WE ARE ALL HEALTHY. THIS IS THE SACRED CIRCLE OF LIFE THAT MUST NOT BE INTERRUPTED.

THE ROOTS OF THE GRASS SHOULD BE SIX FEET DEEP.

BUT THEY HAVE RETREATED. THEY ARE DAMAGED BY ALL THE CHEMICALS WE SPRAY, THE MACHINERY WE USE, THE METHODS WE APPLY.

THE EARTH IS WOUNDED. SHE HAS OPEN SCARS.
BUT LIKE A SCAB WE KEEP PICKING AT THEM...

THE EARTH IS OUR SHARED BODY.

THE RAINFORESTS ARE OUR LUNGS--

--THE CORAL REEFS OUR ARTERIES--

--SO WHY DO WE CONTINUE TO POISON OURSELVES?

FOR SOME REASON WE BELIEVE THAT THE EARTH NEEDS US TO DO HER JOB FOR HER, A JOB THAT SHE HAS ALREADY BEEN DOING FOR CENTURIES.

THE BUFFALO, WITH HIS SPLIT HOOVES, TILLS THE EARTH...

...AND THE HORSE SOWS THE SEEDS, USING HIS TAIL TO BRUSH THEM FROM THE CROPS AND HIS FLAT HOOVES TO PLANT THEM INTO THE EARTH.

THE BUFFALO AND THE HORSE PERFORM THEIR RITUALS TO RESTORE THE EARTH. I'VE SEEN THE BUFFALO DANCE ACROSS THE PLAIN, JUMPING INTO THE AIR LIKE DEER!

THE VIBRATION OF THEIR MOVEMENTS SPEAKS TO THE SOIL. IT TELLS IT IS SAFE TO BE WILD AGAIN.

THE HORSE FOLLOWS THE SAME PATH AS THE BUFFALO. IT IS JUST LIKE A DANCE. THEY DO THE CEREMONY TWO, THREE TIMES EACH...

...AND THE FINAL TIME THEY DO IT TOGETHER.
THE SPIRIT OF THE EARTH IS COMING BACK BECAUSE WE HAVE **GIVEN** IT BACK. THE EARTH KNOWS MORE THAN US. SHE WILL HEAL HERSELF IF WE GIVE HER A CHANCE.

IT IS MY HOPE THAT ONE DAY WE WILL LEARN TO STOP THINKING OF THE EARTH AS OUR RESOURCE, AND INSTEAD AS THE SOURCE TO OUR LIVES, ALL LIFE.
EVERYTHING HAS BEEN GIVEN TO US; IT DOESN'T MEAN WE HAVE TO TAKE IT.

YOU UNDERSTAND, T'UNSHKÁ?

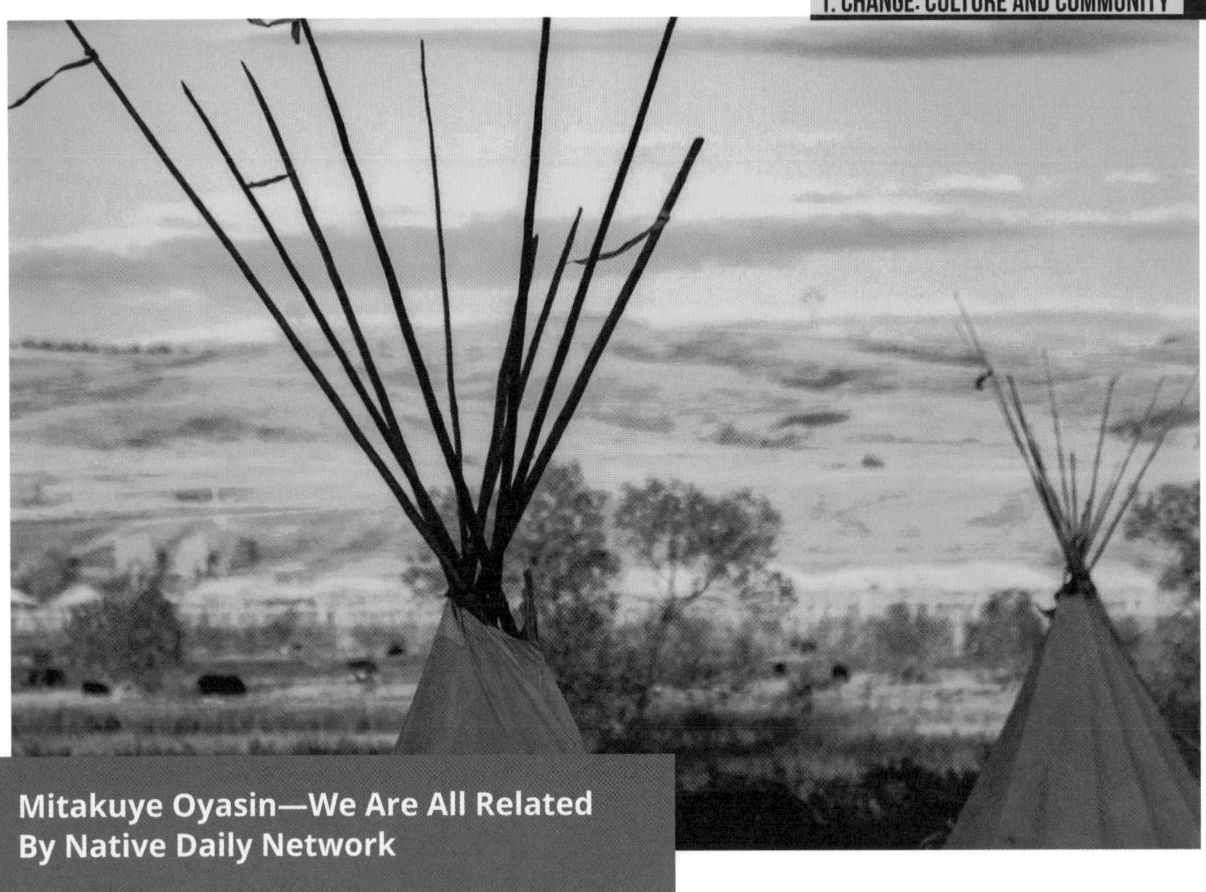

Mitakuye Oyasin—We Are All Related
By Native Daily Network

Traditional Indigenous teachings are an important connection to a time before paradise was lost in all but a few corners of the planet.

CONNECTIONS RUN DEEP.

The Lakota have a saying for this; Mitakuye Oyasin. Roughly translated, it means 'We Are All Related.'

It is an affirmation of a very real connection that exists between all things.

Our interconnected existence is celebrated and cherished. It is a way of life that places us as part of the natural world, not apart from it.

It means that everything is our relative, from the smallest amoeba to the largest whale. From the forests to the water flowing through our rivers. In this way of life, it is understood that each relative has their own, equally important part to play in the natural order.

As Water Protectors we are part of a global community of people that are standing up to protect Unci Maka—Grandmother Earth.

When people ask the Lakota why they fight so hard to protect their waters, they say 'Mni Wiconi'—Water of Life.

WITHOUT IT, THERE IS NO LIFE.
WITHOUT THAT, THERE IS NO US.

BENEATH THE TREES

DEVELOPED WITH	WRITTEN BY	ART & LETTERING BY
ELIZABETH WATHUTI	DAVID BAILLIE	JUNI BA

WHAT ARE THEIR NAMES?

-KZZRT-

THESE TROUBLE MAKERS.

"UNTIL YOU DIG A HOLE, YOU PLANT A TREE, YOU WATER IT AND MAKE IT SURVIVE, YOU HAVEN'T DONE A THING. YOU ARE JUST TALKING."
- WANGARI MAATHAI

ARTHUR GOLONG DID NOT WANT TO BECOME A COMMUNITY LEADER, BUT TYPHOON HAIYAN IN NOVEMBER OF 2013 LEFT HER NO CHOICE.
ONE OF THE DEADLIEST TYPHOONS ON RECORD DEVASTATED HER COMMUNITY IN TACLOBAN CITY.
FOR ISLAND STATES LIKE THE PHILIPPINES, RISING SEAS AND NATURAL DISASTERS, EXACERBATED BY CLIMATE CHANGE, AFFECT THE COMMUNITIES PROFOUNDLY.
"I WAS FOCUSED ON MY WORK AS A HAIRDRESSER. I DIDN'T THINK OF ANYTHING ELSE AS LONG AS I HAD A HOME AND MY FAMILY WITH ME...
ARTHUR'S STORY
DEVELOPED WITH ARTHUR GOLONG
WRITTEN BY GREG LOCKARD
ART & COLORS BY DIIGII DAGUANA
LETTERING BY JIM CAMPBELL
"HOWEVER, WHEN HAIYAN STRUCK, I GOT CONCERNED THAT WE MIGHT NOT BE ABLE TO GET DONATIONS OR AID SINCE NO ONE WAS MANAGING OR REPRESENTING OUR AREA.
"MY NEIGHBORS ENCOURAGED ME TO LEAD BUT I DIDN'T WANT TO BECAUSE I HAVE NO EXPERIENCE. I REALIZED THAT IF NO ONE IS GOING TO STEP UP...WE MIGHT GET LEFT BEHIND.
"I DIDN'T SEE MYSELF AS A LEADER... I JUST DID WHAT WAS NECESSARY.
"I AM FULFILLED WHEN I'M ABLE TO HELP PEOPLE AND SOLVE PROBLEMS FOR THE BENEFIT OF OTHERS."

"NORMALLY IN THE PHILIPPINES, THOSE WHO GET ELECTED AS LEADERS ARE MEN, AND SOMETIMES, WOMEN. NO MEMBERS OF THE LGBTQ COMMUNITY GET ELECTED.
"AS THE FIRST TRANSGENDER WOMAN WHO BECAME A LEADER--FROM A PLACE OF ORIGIN IN OUR COMMUNITY--PEOPLE SAW HOW DIFFERENT MY LEADERSHIP WAS COMPARED TO OTHER RESETTLEMENT SITES.
"SINCE THEN, MORE AND MORE PEOPLE FROM THE LGBTQ COMMUNITY HAVE GROWN INTO LEADERS AS THEY GET ELECTED IN THEIR AREAS.
"THERE IS THAT ACCEPTANCE AND REALIZATION THAT THE LGBTQ ALSO HAVE A SPACE IN LEADERSHIP.
"THIS IS ONE OF THE THINGS I ADVOCATE FOR, SPEAKING UP ABOUT THE NEEDS OF THE LGBTQ COMMUNITY ON TACKLING DISCRIMINATION AND PROMOTING INCLUSIVITY IN LIVELIHOOD PROGRAMS OFFERED AFTER HAIYAN.
"THE WORLD NEEDS TO KNOW THAT THE LGBTQ SECTOR IS THE MOST VULNERABLE BECAUSE UNTIL NOW, THERE IS STILL DISCRIMINATION, ESPECIALLY IN THE CONTEXT OF OPPORTUNITIES. WE ARE ALWAYS LEFT BEHIND WHEN IT COMES TO EMPLOYMENT OR LIVELIHOOD OPPORTUNITIES."

"I FIND MOMENTS OF PEACE IN OUR FRONT YARD. IT'S WHERE MY PLANTS ARE AND WHERE I GO EVERY MORNING AND AFTERNOON. I FEEL AT PEACE KNOWING THAT PEOPLE WHO TRUSTED ME THEN STILL HAVE CONFIDENCE IN ME AND CONTINUE TO SUPPORT ME NOW.
"BEFORE, MY SOURCE OF MOTIVATION WERE MY PARENTS, BUT BOTH OF THEM PASSED AWAY LAST YEAR. THEY WERE MY INSPIRATION. NOW, I ENJOY THE COMPANY OF MY NEPHEWS AND NIECES WHO ARE ALWAYS HERE.
"OF COURSE, I AM HOPEFUL THE GLOBAL COMMUNITY CAN CHANGE THEIR ATTITUDES TOWARD CLIMATE CHANGE.
"FOR ME, WE SHOULD NOT BE DRIVEN BY PRIDE OR EGO. WE SHOULD BE DANCING TO THE SAME BEAT. IF WE WANT TO ADDRESS THE PROBLEM OF CLIMATE CHANGE, GLOBALLY WE MUST BE UNITED. BECAUSE LET'S FACE IT, WE ALL CONTRIBUTE TO POLLUTION SO WE ALL MUST WORK TOGETHER TO FIND A SOLUTION...
"I'LL LOOK FORWARD TO THAT DAY."

It's freezing cold.

(I could be at home...)

GREENPEACE CLIMATE CHANGE PROTEST.

A BP OIL RIG, STATIONED IN THE CROMARTY FIRTH.

DEPTHS

DEVELOPED WITH	WRITTEN BY	ART BY	COLORS BY	LETTERING BY
MEENA RAJPUT	COREY BROTHERSON	GERALDO BORGES	LEE LOUGHRIDGE	BERNARDO BRICE

Poisonous words from toxic minds.
And through it all, I hear their voices again. Trying to pull me down.
You can't change things, Meena. You're just a girl.
There's no future in environmental careers.
[vile racist epithet]
Their world is so small.
We need to be *bigger.*
Their voices didn't stop me *then.* And this cold won't stop me *now.*
Apathy *isn't* strength.
And as a child I realized even a single drop can create waves.
Others helped show me that. So I pay that forward.
Standing up to bullying.
Empowering others.
And every time doubters say: *"But you can't..."*
I simply shake my head.
"I can."

I've been asked "Can't someone else do it?"
No. No matter the cold. No matter the difficulty. No.
Stop looking away--

--and start looking up.
Look at *everything* around us. Our society. Our environment. *Us.*
Change only starts and ends with us.
My family always says: "Bad things never happen when you try."
CLIMATE EMERGENCY
So I'm always going to keep trying.

On 14th June 2019, Meena and her Greenpeace UK co-protester Andrew climbed and chained themselves to the Transocean/BP oil rig, Paul B Loyd Junior. Between them and six others, they occupied the rig for a combined total of almost six days.

Their actions prevented the rig from leaving the North Sea to the Vorlich oil field, helping raise awareness of the injustices and dangers of fossil fuel drilling during a mounting climate emergency.

Meena continues to work at Greenpeace as an activist, while also focusing on establishing an anti-racist culture through action and her own children's books.

COMMUNITY AND CULTURE

BY MEENA RAJPUT

Think about it like this: Would governments and corporations still drag their heels with climate solutions if it were their richest citizens who were suffering the consequences of climate change? Hurricanes, droughts, flooding on a daily basis? NO, they wouldn't.

Yet hundreds of thousands of people of colour and Indigenous communities around the world are experiencing, death, displacement and suffering caused by environmental destruction and climate related disasters.

Despite knowing this, governments and corporations do very little to change their policies and business models - business models that directly cause climate change and are built on histories of human exploitation. They know it must stop, but they continue driven by greed, while silencing the solutions proposed by impacted communities. As well as causing social injustice, they have disrupted the climate and nature to the point of threatening humanity's existence.

We all have a choice. We can continue accepting the way things are or we fight for change.

Change will come if we all suppport and join communities fighting for a better world, if communities unite with one another, and if we promote a culture of justice and equality.

WHAT YOU CAN DO TO HELP:

1. Connect and act in solidarity with communities fighting for change.
2. Think of ways you can tackle systemic inequality and the climate crisis.
3. Take action together and influence leaders to drive change.

A CLIMATE OF HOPE

DEVELOPED WITH	ADAPTED BY	ART & COLORS BY	LETTERING BY
PROF. MICHAEL MANN	ROGER STERN	JOE ORSAK	JIM CAMPBELL

TRASH
PLEASE RECYCLE
"IT IS ALL OF THE THINGS WE HAVE TALKED ABOUT--

"BEHAVIORAL CHANGE,
PUT CARBON TO GOOD USE
PLANT A TREE
JORSAK

"INCENTIVIZED BY APPROPRIATE GOVERNMENT POLICY,

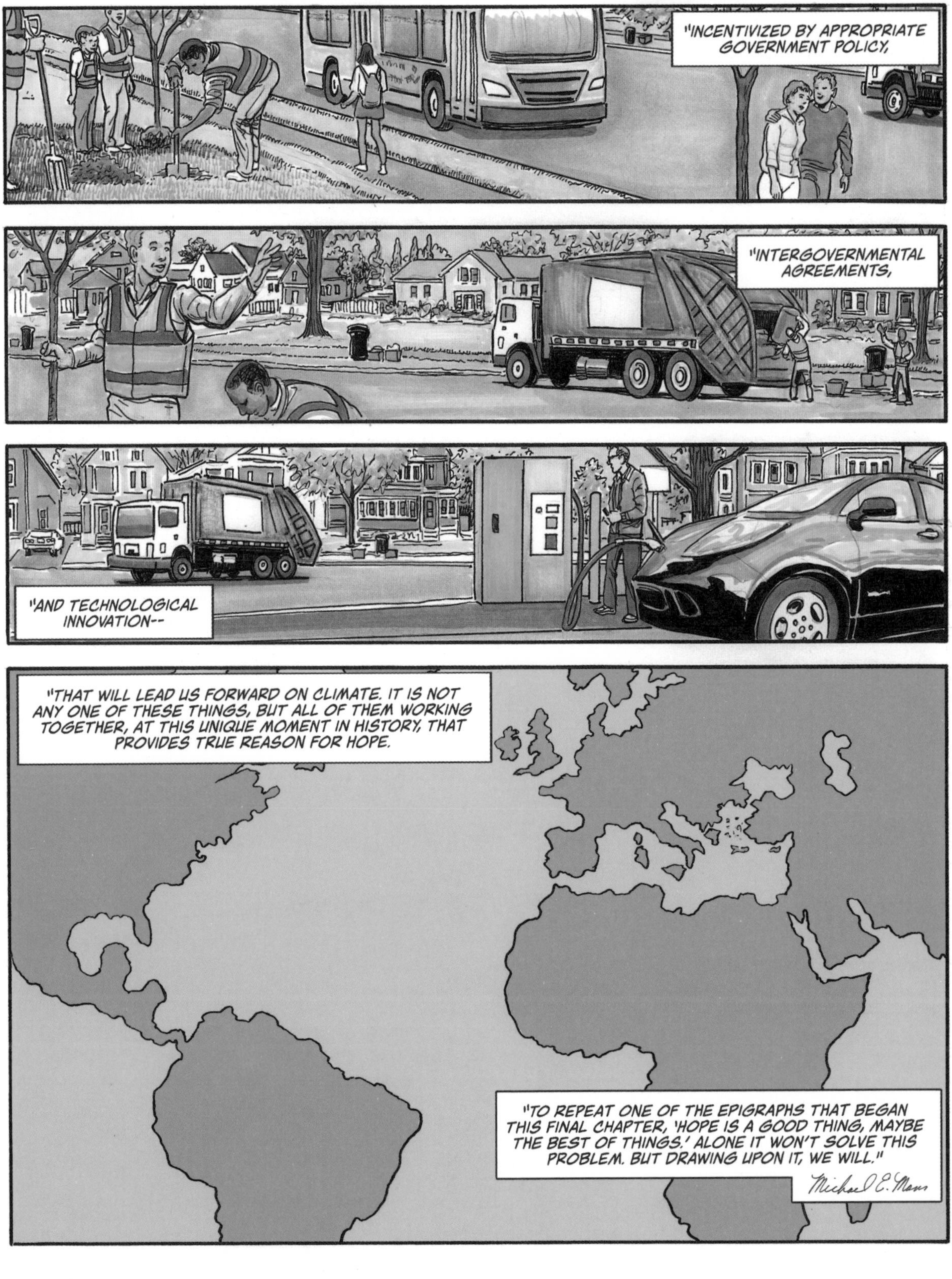

"INTERGOVERNMENTAL AGREEMENTS,

"AND TECHNOLOGICAL INNOVATION--

"THAT WILL LEAD US FORWARD ON CLIMATE. IT IS NOT ANY ONE OF THESE THINGS, BUT ALL OF THEM WORKING TOGETHER, AT THIS UNIQUE MOMENT IN HISTORY, THAT PROVIDES TRUE REASON FOR HOPE.

"TO REPEAT ONE OF THE EPIGRAPHS THAT BEGAN THIS FINAL CHAPTER, 'HOPE IS A GOOD THING, MAYBE THE BEST OF THINGS.' ALONE IT WON'T SOLVE THIS PROBLEM. BUT DRAWING UPON IT, WE WILL."

Michael E. Mann

£1,000,000
Help fund our global oceans campaign for a year.

£500,000
Support world-class scientific research on wildlife ecosystems.

£100,000
Enable us to lobby decision-makers to agree to a Global Oceans Treaty.

£50,000
Provide resources to investigate destructive fishing practices.

GREENPEACE

THE MONEY WE RAISE IS HELPING:

SECURE GLOBAL OCEAN PROTECTION

The oceans are home to incredible wildlife, provide food for billions and help balance our climate; yet only a tiny fraction is protected. For decades humans have plundered and polluted this precious environment. We urgently need to establish a global network of ocean sanctuaries to help our oceans recover.

WHY IS THE PROJECT SO IMPORTANT?

Our fate is bound to the fate of our oceans. They do more than harbor incredible wildlife—they support all life on Earth. By producing half our oxygen and soaking up huge amounts of carbon dioxide, oceans are one of our best defenses against climate change. If they don't make it, we don't, either.

Protecting our oceans needs international cooperation. Greenpeace is campaigning around the world to persuade governments to agree to a strong Global Ocean Treaty at the United Nations. This is an historic opportunity to create ocean sanctuaries, areas safe from human exploitation, across at least 30% of our oceans by 2030.

Wherever a proper ocean sanctuary is created, the results are dramatic. Habitats recover. The fish come back. Life finds a way. A global network of sanctuaries is a brilliantly simple—and achievable—solution to some of the threats our oceans face.

SPECIES PROFILE

Every year thousands of turtles are caught in industrial fishing nets and their feeding grounds and breeding habits are threatened by climate change. Most species face extinction. Protected areas will give sea turtles and other marine life the chance they need to recover and thrive.

> *We need to protect the ocean as if our lives depend on it, because they do.*
>
> Dr. Sylvia A. Earle, marine biologist, explorer, and author

BURU SEA, INDONESIA
Handline fishing on Buru Island.

PROJECT IN ACTION

This project will inspire millions around the world to add their voices to the campaign, such as during this World Oceans Day Event in Mexico. It will support leading scientists to deliver new research, making the case for global oceans protection. And it will put pressure on decision-makers at the highest levels of government.

THE FUTURE

If we succeed, this global network of ocean sanctuaries will be one of the biggest conservation efforts in human history, creating millions of square kilometers of new protected areas. Ocean life would become healthier, more abundant, and better able to cope with big global threats like climate change and plastic pollution.

FIND OUT MORE

SIGN THE PETITION: HTTP://GREENPEACE.ORG.UK/PROTECT-THE-OCEANS
FIND OUT MORE: HTTPS://WWW.GREENPEACE.ORG.UK/CHALLENGES/OCEAN-SANCTUARIES/

CHAPTER TWO

PROTECT

THE WORLD

Currently, our most precious and biodiverse areas are under attack. From business exploitation to fishing waste, mining to illegal wildlife trade, the threats are numerous but the solution is simple...

Often this land is occupied by Indigenous communities who daily face the harsh realities of external commercial land-use change on the delicate ecosystems they have respected for generations. And when these incredible pockets of paradise are gone, we all suffer.

We are working with World Land Trust to buy some of the most important land on Earth, and gift it back to the indigenous communities who live there, safeguarding it for those who have proven time and time again to be the best stewards of it.

Sponsored project:

THE SHEPHERD & THE THUNDER

THERE IS A PROVISION IN THE UNITED NATIONS WORLD CHARTER FOR NATURE THAT CALLS ON NONGOVERNMENTAL GROUPS TO ASSIST IN SAFEGUARDING NATURE IN AREAS BEYOND NATIONAL JURISDICTION.
DEVELOPED WITH
PETER HAMMARSTEDT
WRITTEN BY
BRIAN AZZARELLO
ART & COLORS BY
DANIJEL ZEZELJ
LETTERING BY
BERNARDO BRICE
THE THUNDER.
GLOBALLY THE MOST NOTORIOUS PIRATE FISHING VESSEL, WAS SINKING OFF THE COAST OF NIGERIA.
MOST LIKELY SCUTTLED BY ITS OWN CREW.
IN THE PAST DECADE, THE THUNDER HAD COLLECTED OVER $76 MILLION FROM ILLICIT SALES OF TOOTHFISH--AKA CHILEAN SEA BASS.
THE THUNDER HAD BEEN BANNED FROM FISHING SINCE 2006.
FOR SEA SHEPHERD CAPTAIN PETER HAMMARSTEDT, A MAN WHO'D DOGGEDLY CHASED THE THUNDER FOR ONE HUNDRED AND TEN DAYS THROUGH TWO SEAS AND THREE OCEANS, THE ENDING WAS BITTERSWEET.
PETER IS A MEMBER OF SEA SHEPHERD, A MARINE CONSERVATION ORGANIZATION THAT ENGAGES IN DIRECT ACTIONS TO DEFEND WILDLIFE AND PROTECT THE OCEANS FROM ILLEGAL EXPLOITATION.
SEA SHEPHERD DESCRIBES ITSELF AS AN ECO-VIGILANTE GROUP.

HE'D IMAGINED IT DIFFERENTLY. BRINGING THE THUNDER TO PORT, TO STAND TRIAL FOR CRIMES, AND A BLATANT DISREGARD FOR THE WORLD'S BIODIVERSITY.
ILLEGAL FISHING GENERATES $100 BILLION ANNUALLY.
BUT THE ACTUAL COSTS TO LIFE IS MUCH HIGHER.
IN THEIR GREEDY PURSUITS, ILLEGAL NETS WERE USED-- NETS THAT DIDN'T JUST HAUL IN "TODAY'S CATCH", BUT **BYCATCH** AS WELL.
"BYCATCH" BEING A BLOODLESS TERM FOR A FISH OR OTHER MARINE SPECIES THAT IS CAUGHT UNINTENTIONALLY WHILE CATCHING CERTAIN TARGET SPECIES.
IT WASN'T JUST SEA-LIFE THAT WAS IMPACTED, EITHER. LOCAL FISHERMEN WERE BEING FORCED FURTHER FROM SHORE BY ILLEGAL OVERFISHING. THE WATERS WERE **DANGEROUS...**
AND SOMETIMES **WORSE.**
SEA SHEPHE
BUT IN THE GABONESE PORT, CAPTAIN HAMMARSTEDT MET LIKE-MINDED INDIVIDUALS. A GOVERNMENT THAT HAD THE WAYS, BUT NOT THE MEANS, TO COMBAT MARINE POACHING.
ALL THEY NEEDED WAS A **BOAT.**

SEA SHEPHERD PROVIDES THE SHIPS, THE CREW, AND THE FUEL.

GOVERNMENT PARTNERS PROVIDE THE LAW ENFORCEMENT TO BOARD, INSPECT, AND ARREST ANY VIOLATORS OF THE LAW.

SO, ONE ENDING LEADS TO A BEGINNING. BUT THAT'S WHAT **PASSION** IS, ISN'T IT?

SEA SHEPHERD

THE THUNDER MAY HAVE **SUNK**, BUT THE BOB BARKER* IS STILL ON **PATROL**.

BECAUSE DEEP SEA SHARK POPULATIONS HAVE BEEN DECIMATED BY 80%.

BECAUSE THE DEATHS OF 300,000 WHALES AND DOLPHINS PER YEAR IS DUE TO BYCATCH.

BECAUSE OVER 50% OF THE WORLD'S MAJOR FISHING GROUNDS HAVE BEEN DEPLETED.

SEA SHEPHERD HAS ASSISTED IN ARRESTING 65 ILLEGAL POACHING VESSELS OFF THE COAST OF AFRICA.

BOB BARKER

* The Bob Barker is named after the former host of ***The Price Is Right***, who generously supported their cause.

To learn what you can do to help, or discover more about the Sea Shepherd's mission, visit www.seashepherd.org

HURRY UP! THEY'RE GETTING AWAY!
THE SURFING HIPPOS OF LOANGE
WHY ARE THEY SHOOTING AT US?
THEY'RE POACHERS! RUN, BABY! RUN!
DEVELOPED WITH
PETER HAMMARSTEDT
WRITTEN BY
ERIKA LEWIS
ART BY
ALTI FIRMANSYAH
COLORS BY
YENNY LAUD
LETTERING BY
BERNARDO BRICE
NO HUMANS
NO POACHING
PEOPLE FREE ZONE
STAY OUT

CHE-CHECK!
CHE-CHECK!
BRACONNIERS!
CHE-CHECK!
CHE-CHECK!
CHE-CHECK!
CHE-CHECK!
SARGENT PEPPER!
MAMMA, YOU HEAR THAT?
KEEP GOING! RUN! ***RUN!***
CHE-CHECK!
DID YOU SEE THOSE SIGNS? I DON'T THINK WE'RE SUPPOSED TO BE IN HERE...
WILL YOU SHUT UP? WHO'S GOING STOP US?
TURN AROUND HUMANS
DANGER
FINAL WARNING
MAMMA! THE WATER IS COMING IN! WHAT DO WE DO?
STAY BEHIND ME, YOU HEAR ME?
NOWHERE TO RUN NOW...

GASP
OOF!
PAFF
CAN'T YOU READ?
AND THEY THINK THEY'RE THE MORE INTELLIGENT SPECIES. YOU KNOW, IF WE WERE RUNNING THINGS—
HOLD UP. I SAW THAT MOVIE. DIDN'T TURN OUT SO WELL....
HUMANS WROTE THAT MOVIE...CAN YOU SAY *PROPAGANDA?*

BABY!
WHERE ARE YOU?
IT'S ALL GOOD. SURF'S UP, MAMMA.
COWABUNGA!
YOU'RE MIGHTY PRETTY, MAMMA.
THANK YOU...
WELCOME HOME... YOU'LL BE SAFE HERE.

SAIKOU'S NIGHTMARE AND A DREAM
KARTONG, GAMBIA.
AFTER A WHOLE DAY FISHING, IT'S ALWAYS GOOD TO BE HOME.
DEVELOPED WITH
MUSTAPHA MANNEH & SAINEY GIBBA
WRITTEN AND PENCILLED BY
ABDULKAREEM BABA AMINU
INKS AND COLORS BY
RYAN CODY
LETTERING BY
BERNARDO BRICE
REST, DINNER, AND AISHA: HOME IS WHERE EVERYTHING IS. WELL, ALMOST EVERYTHING.

WELCOME HOME, SAIKOU. LET ME TAKE YOUR BAG.
THANK YOU, DEAR WIFE. YOU'RE FAR TOO KIND.

I'M AFRAID THERE'S NO FISH TODAY EITHER, MY DEAR...
DON'T BE SILLY. I JUST WANT TO AIR OUT YOUR WORK CLOTHES.

GOOD NIGHT.
SLEEP WELL.

I SLEEP...
...AND IT'S THE SAME NIGHTMARE AS THE NIGHT BEFORE.

Gambia has three fishmeal factories operating on the coastline communities. All three are Chinese owned, and their practices have destroyed local fishing trade and resulted in the pollution of our local waters and fish.

They process massive amounts of raw fish into fishmeal and oil, the scale of which is causing a national disaster for the people of Gambia.

ANOTHER NIGHTMARE?
YES.

SADLY, MY NIGHTMARE IS ALSO OUR REALITY.

BUT LIKE EVERY OTHER DAY OR NIGHT, I HAVE HOPE THAT THINGS WILL GET BETTER TODAY.

SAIKOU'S NIGHTMARE – AND HIS DREAM – AREN'T HIS ALONE.

THEY'RE OURS.

WHAT DID YOU WISH FOR?
HETTIE HAD JUST TURNED THIRTEEN WHEN HER PARENTS SENT HER OFF TO SAILING CAMP.

AT FIRST, THE LOGIC OF SAILING ELUDED HER.
WHAT HAPPENS WHEN THE WIND IS NOT BLOWING WHERE YOU WANT TO GO?
THE WIND IS INVISIBLE, BUT HER EFFECTS ARE NOT.
IF YOU UNDERSTAND HER STRENGTH AND HER DIRECTION, SHE CAN TAKE YOU ANYWHERE...

...A SAILOR WAS BORN.

IN 1991, MOUNT PINATUBO ERUPTED, MAKING MIGRANTS OF MANY IN THE PHILIPPINES.

THE U.S. MILITARY ABANDONED THEIR MILITARY BASES, LEAVING BEHIND A SHELTER AND A TOXIC LEGACY.
BUT LIFE BROKE THROUGH IN THAT SPACE...

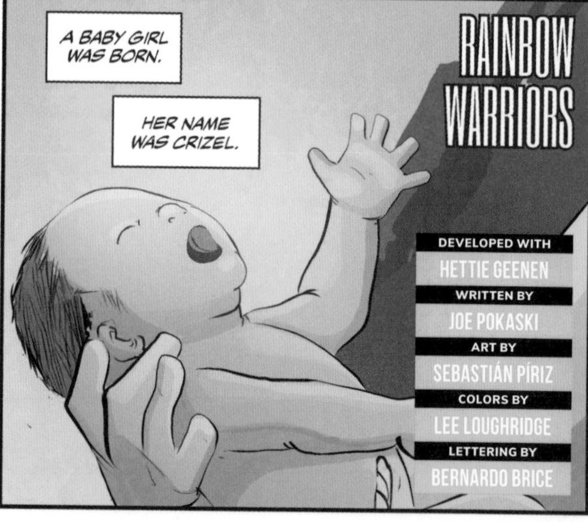
A BABY GIRL WAS BORN.
HER NAME WAS CRIZEL.
RAINBOW WARRIORS
DEVELOPED WITH
HETTIE GEENEN
WRITTEN BY
JOE POKASKI
ART BY
SEBASTIÁN PÍRIZ
COLORS BY
LEE LOUGHRIDGE
LETTERING BY
BERNARDO BRICE

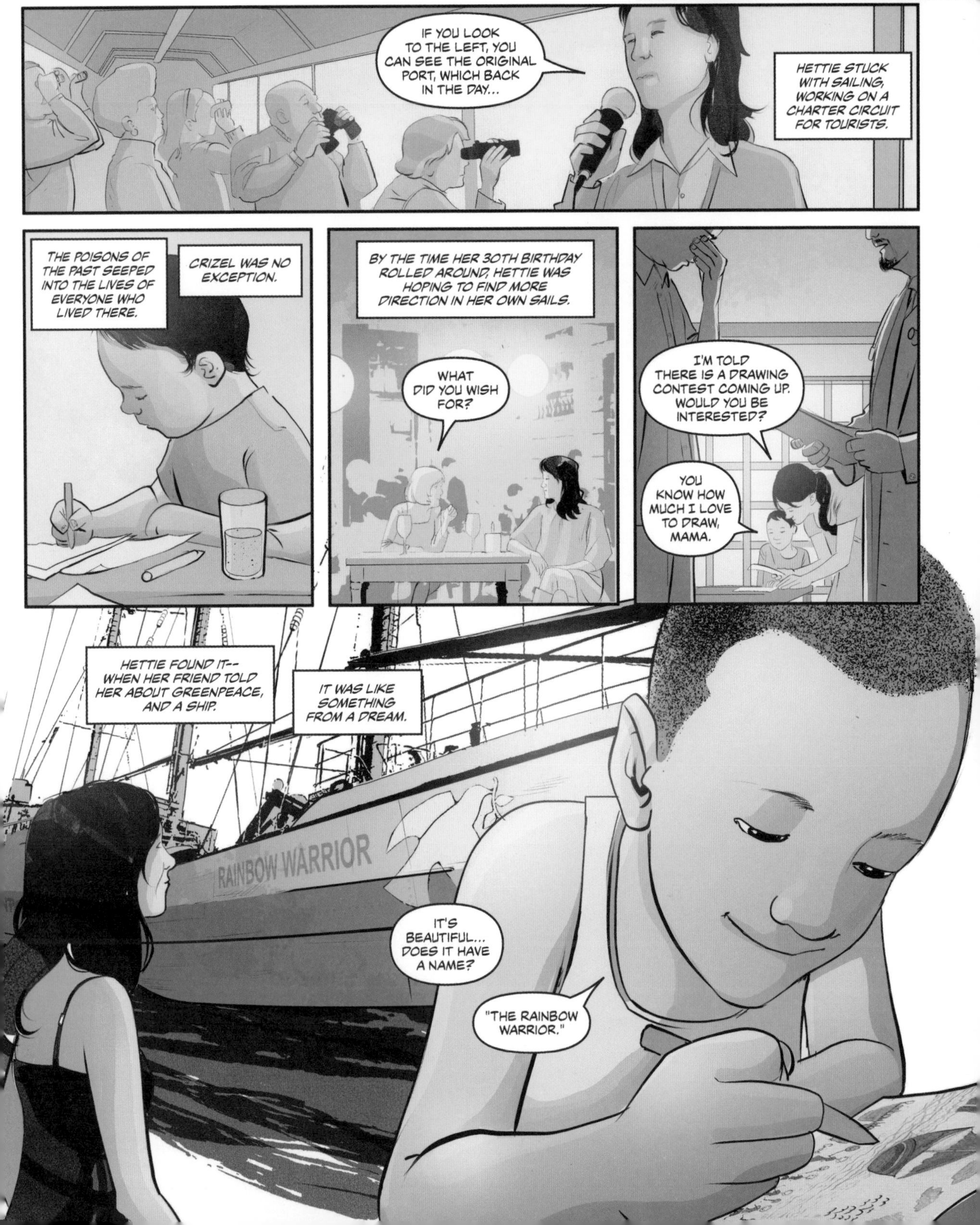

IF YOU LOOK TO THE LEFT, YOU CAN SEE THE ORIGINAL PORT, WHICH BACK IN THE DAY...
HETTIE STUCK WITH SAILING, WORKING ON A CHARTER CIRCUIT FOR TOURISTS.
THE POISONS OF THE PAST SEEPED INTO THE LIVES OF EVERYONE WHO LIVED THERE.
CRIZEL WAS NO EXCEPTION.
BY THE TIME HER 30TH BIRTHDAY ROLLED AROUND, HETTIE WAS HOPING TO FIND MORE DIRECTION IN HER OWN SAILS.
WHAT DID YOU WISH FOR?
I'M TOLD THERE IS A DRAWING CONTEST COMING UP. WOULD YOU BE INTERESTED?
YOU KNOW HOW MUCH I LOVE TO DRAW, MAMA.
HETTIE FOUND IT-- WHEN HER FRIEND TOLD HER ABOUT GREENPEACE, AND A SHIP.
IT WAS LIKE SOMETHING FROM A DREAM.
RAINBOW WARRIOR
IT'S BEAUTIFUL... DOES IT HAVE A NAME?
"THE RAINBOW WARRIOR."

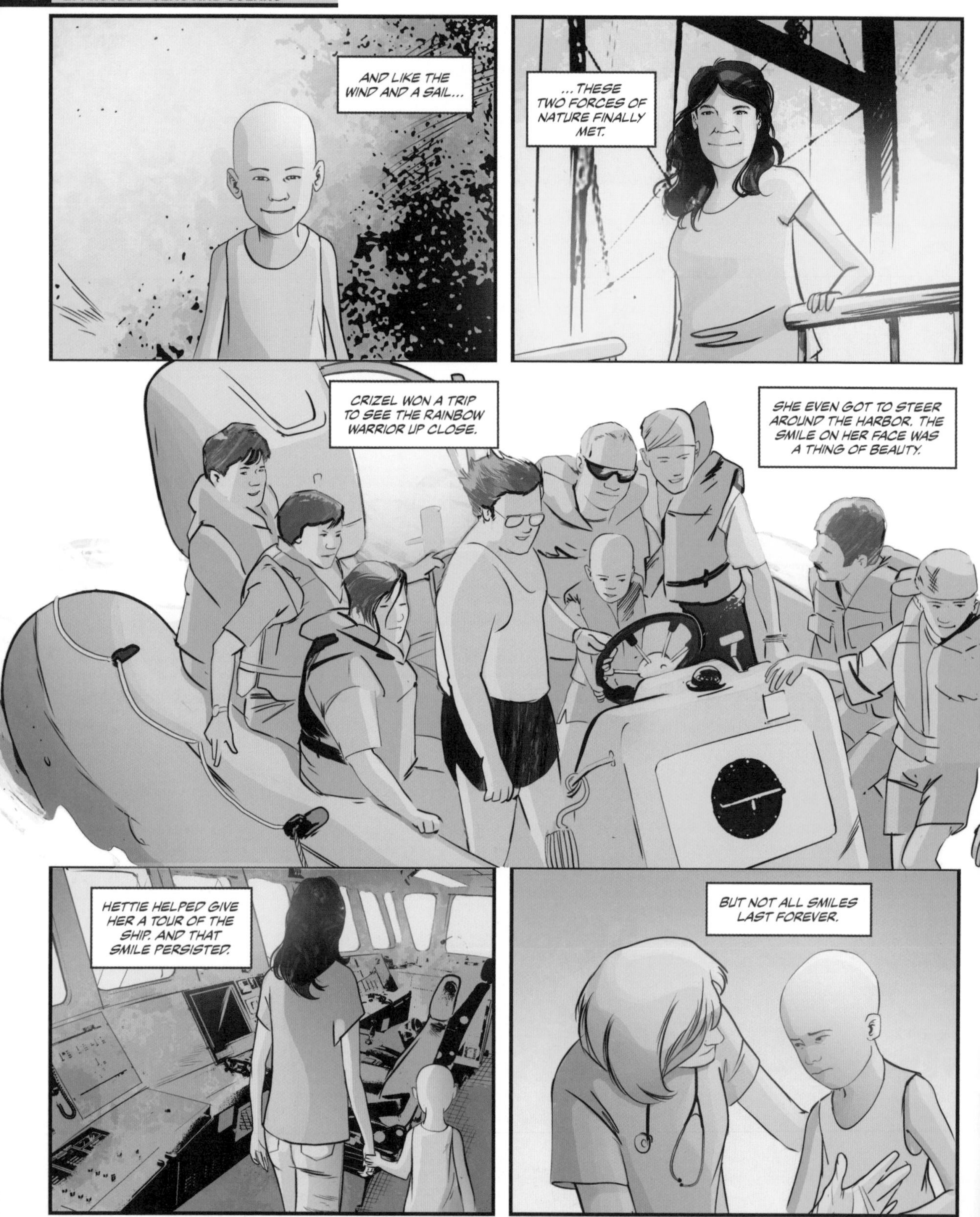
AND LIKE THE WIND AND A SAIL...
...THESE TWO FORCES OF NATURE FINALLY MET.
CRIZEL WON A TRIP TO SEE THE RAINBOW WARRIOR UP CLOSE.
SHE EVEN GOT TO STEER AROUND THE HARBOR. THE SMILE ON HER FACE WAS A THING OF BEAUTY.
HETTIE HELPED GIVE HER A TOUR OF THE SHIP. AND THAT SMILE PERSISTED.
BUT NOT ALL SMILES LAST FOREVER.

IT WAS CRIZEL'S DREAM, TO SEE THE BOAT, TO UNDERSTAND WHAT HOPE LOOKED LIKE. TO HOPE IT COULD SAVE THE WORLD.

AND IT WAS SEEING THIS SHIP THROUGH THIS YOUNG GIRL'S EYES THAT HELPED HETTIE SEE HER LIFE MORE CLEARLY.

ARE YOU GOING TO MAKE A WISH?
"THE WIND IS INVISIBLE, BUT HER EFFECTS ARE NOT.
"IF YOU UNDERSTAND HER STRENGTH AND HER DIRECTION, SHE CAN TAKE YOU ANYWHERE..."
THE END

WRITTEN AND DRAWN BY
VICTOR SOLÍS

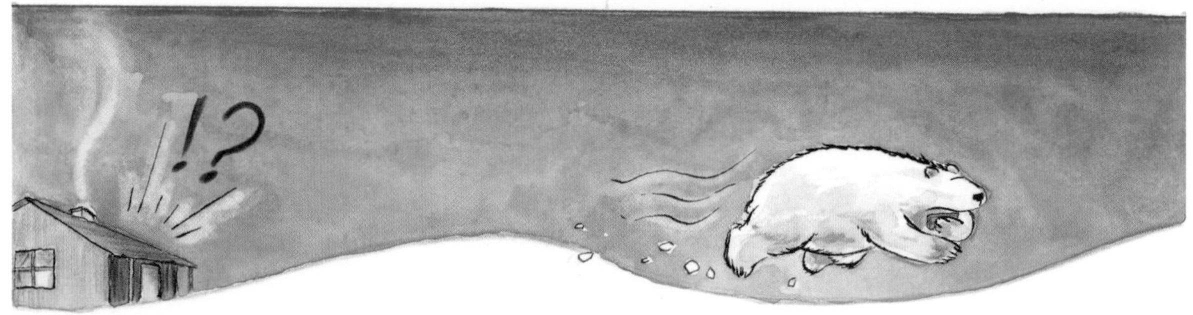

THE SELFLESS GIANT

DEVELOPED WITH	WRITTEN BY	ART & COLORS BY	LETTERING BY
ANDY SERKIS	PAUL GOODENOUGH	DANIJEL ZEZELJ	BERNARDO BRICE

SKAFTAFELLSJOKULL, ICELAND.

My granddad died yesterday...

WE ARE LOSING ARTIC ICE AT THE RATE OF 13% PER DECADE.
IF WE CONTINUE AT THIS RATE, EARTH COULD BE ICE FREE BY 2040.

THIS STORY IS DEDICATED TO THE SELFLESS GIANT GLACIERS THAT KEPT US SAFE FOR SO LONG.
WE ARE SORRY. WE WILL DO BETTER. WE HAVE TO.

School's Out

War and Peas

WRITTEN BY	ART BY
TAIKA WAITITI, PAUL GOODENOUGH AND SAFELY ENDANGERED	SAFELY ENDANGERED

SHO' IS LONELY BEIN' ALL BY MY LONESOME.
OHNO
© 2021 AZZ + CHIANG (AFTER KELLY)
WRITTEN BY BRIAN AZZARELLO
ART & LETTERING BY CLIFF CHIANG

I BEEN THINKIN'--
I THOUGHT I SMELLED SOMETHIN' BURNIN'.

THAT MUST BE THE WILDFIRES.
THEM AGIN? ALL I'M SMELLIN' IS A SEA LEVEL ROSE.

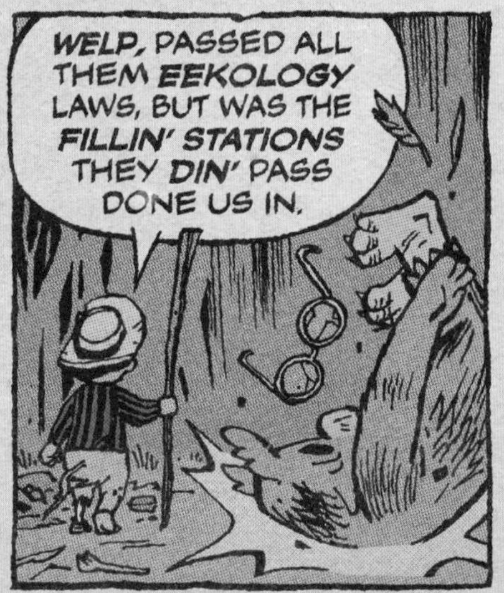
WELP, PASSED ALL THEM EEKOLOGY LAWS, BUT WAS THE FILLIN' STATIONS THEY DIN' PASS DONE US IN.

FOSSIL FUEL, REPORTIN' FER DUTY, SIR.
GOOD TO HAVE YOU, MY PRIVATES, FOR THE WAR IS LOST.

WHAT'S EATIN' YA, OHNO?
AIN' YOU, GATOR.
MINE

WELL, NO, I LOST ME MY APPLETIGHT, ONCE I CAUGHT ME THE STINKSHUN.

YOU CAUGHT THE STINKSHUN? HOW LONG WAS YOU DOWN WITH IT?
FER THE COUNT, FRIEN'. DOWN FER THE COUNT.
MINE

TWIXT THE OCEANS AN' NFERNOS, OKEFENOKEE URLED 'TWEEN THE WAVES.
AW, YOU MAKE IT SOUN' SO NICE.

SO SAD! DEM SCIENCES YOU SPOUTIN' JUS' FAKE NEWS! THIS HERE IS JUS' MOTHER NATURE, NOT A LACK OF HUMAN NURTURE!
TWEET TWEET
TWEET

WHAT?!?
YES...

WE HAVE MET THE ENEMY, AND THEY IS YOU.

Important Takeaway

WRITTEN BY STEVE BACKSHALL
ART & LETTERING BY HARRY VENNING

BY PETER HAMMARSTEDT

Sea Shepherd Captain of the *Bob Barker*. Illustration by Steve White

Over fifty years ago, the French ocean explorer Jacques Cousteau said that:

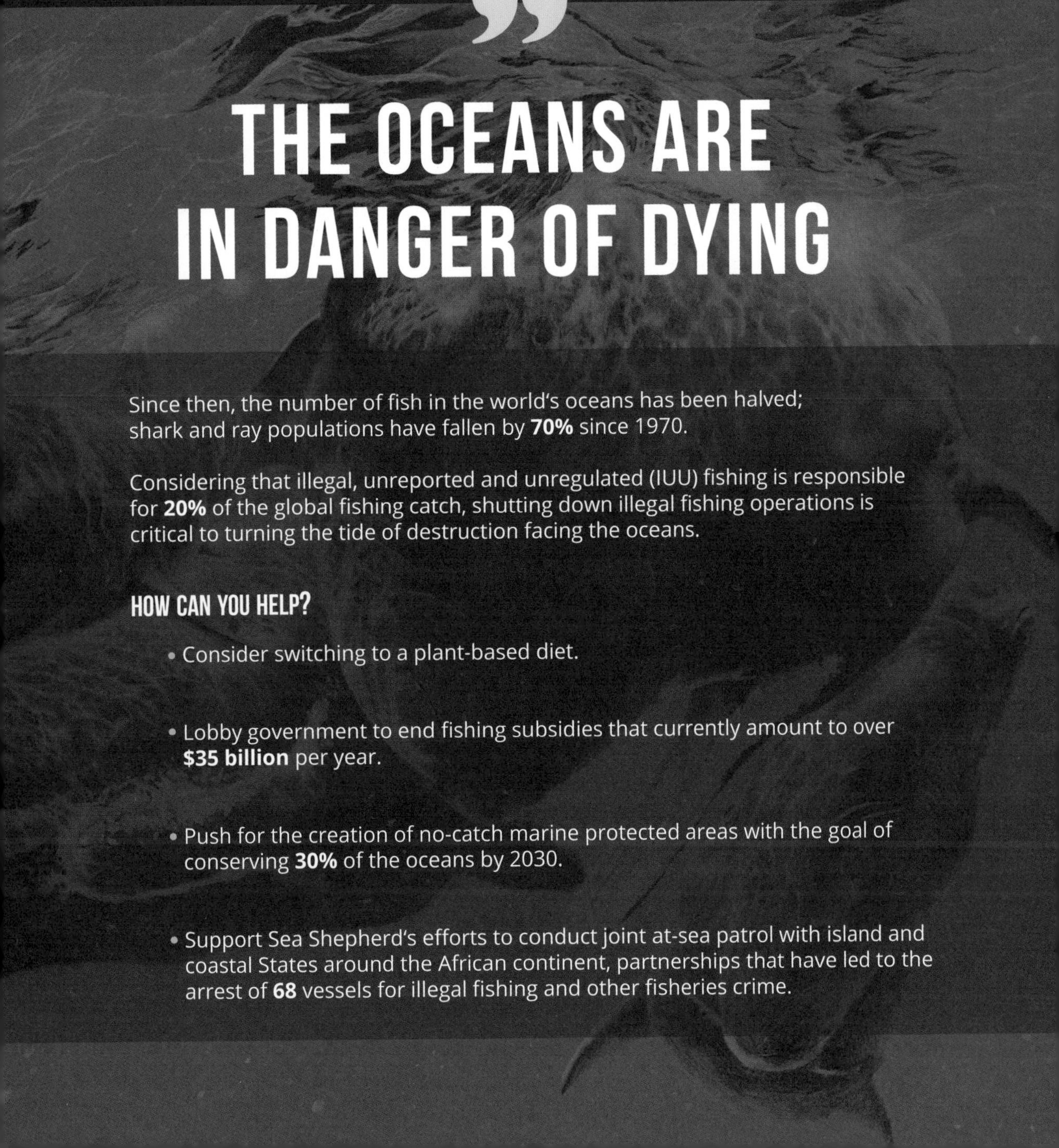

"THE OCEANS ARE IN DANGER OF DYING

Since then, the number of fish in the world's oceans has been halved; shark and ray populations have fallen by **70%** since 1970.

Considering that illegal, unreported and unregulated (IUU) fishing is responsible for **20%** of the global fishing catch, shutting down illegal fishing operations is critical to turning the tide of destruction facing the oceans.

HOW CAN YOU HELP?

- Consider switching to a plant-based diet.
- Lobby government to end fishing subsidies that currently amount to over **$35 billion** per year.
- Push for the creation of no-catch marine protected areas with the goal of conserving **30%** of the oceans by 2030.
- Support Sea Shepherd's efforts to conduct joint at-sea patrol with island and coastal States around the African continent, partnerships that have led to the arrest of **68** vessels for illegal fishing and other fisheries crime.

WRITTEN BY
DAVID SCHNEIDER, AMBER WEEDON AND SAFELY ENDANGERED

ART BY
SAFELY ENDANGERED

AS THE RHYTHM SWELLS, WE...
... RISE.
WE SENSE THE THREAT. FEEL THE NEED...
... FOR EACH OTHER'S STRENGTH.
FOR
JENGI
DEVELOPED WITH
JONATHAN BARNARD
WRITTEN BY
SIMON FURMAN
ART & LETTERING BY
TIMO WUERTZ

UNITED, OF ONE MIND AND PURPOSE...

... WE LOCATE THE SOURCE OF THE DANGER.

THE RED X SPELLS DEATH.

THE JENGI EXPLAINS IN THE DREAM-TONGUE...

SEE - THE TRAIL OF DEVASTATION WROUGHT BY THE INVADERS.

SEE - THE CALAMITOUS FOOTPRINT OF OUR ENEMY.

SEE - A WORLD PUSHED TO THE BRINK.

BUT IN UNITY THERE IS DEFIANCE.

IN TOTALITY - RESISTANCE.
A MILLION SMALL ACTIONS... FROM SMALLEST TO LARGEST...
CAN SAVE US!
THIS... IS JENGI - THE SPIRIT OF THE FOREST. DIRECTED, CHANNELED.
WE HEAR THE RHYTHM. IT IS A CRESCENDO OF HOPE.
TOGETHER WE RISE!

TROPICAL FORESTS CONTAIN AT LEAST TWO-THIRDS OF EARTH'S BIODIVERSITY IN LESS THAN 10% OF ITS SURFACE AREA.

BY WORLD LAND TRUST AND RESERVA: THE YOUTH LAND TRUST
ILLUSTRATION BY NEIL BLACKBIRD SIMS

Home to hundreds of thousands of known species and millions more yet to be discovered or described by scientists, each patch of forest is as unique as the human fingerprint; once an area is lost, it can never be truly recovered. 28,400 acres, or 45 square miles, are lost every day.

In place of these botanical marvels, we install industrial-scale plantations of soy and palm oil, pastures for cattle, and mining operations of metals, oil, gas and timber.

But it's not too late.

The best tool to fight the combined threats of climate change and biodiversity loss is the conservation of tropical forests. As world leaders work on global goals to conserve 30% of the planet by 2030, what you do makes a difference.

1. REDUCE DEFORESTATION THROUGH EVERYDAY CHOICES

Eat locally sourced food, shop antiques before ordering new furniture, eat a plant-rich diet, divest from gold and oil, and get in the habit of asking the question, "Where did this come from?"

2. RESTORE LOCAL FORESTS

You may not live in the Amazon Rainforest, but your local ecosystem is almost certainly in need of help.
Volunteer with a local conservancy to plant native trees and remove invasive species.

3. PROTECT WHAT REMAINS

Organize fundraisers to protect habitat through organizations like the ones highlighted here.
Whether you're an artist, a teacher, or you know some good jokes, you have skills that can be converted into acres of protected forest through creative fundraising.

TOGETHER WE CAN ENSURE THE WORLD'S TROPICAL PRIMARY FORESTS CONTINUE TO STAND TALL.

SAVE LAND. SAVE SPECIES. SAVE FORESTS.

There's Blood On Your Hands
DEVELOPED WITH
CALLIE BROADDUS & JAVIER ROBAYO
WRITTEN BY
TYLER JENNES
ART & COLORS BY
ANEKE MURILLENEM
LETTERING BY
JIM CAMPBELL
SAVE THE PLANET
DING DING
EXCUSE ME, MISS?
YOU HAVE BLOOD ON YOUR HANDS.

CHOCÓ RAINFOREST,
ECUADOR.
16
FRAGILE

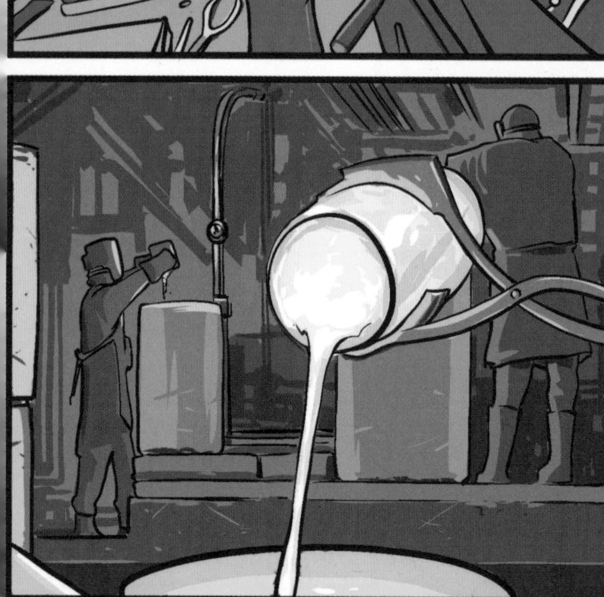

WITH BIODIVERSIFICATION AND RESTORATION EFFORTS GAINING MORE PUBLIC ATTENTION, THE CHOCÓ NEEDS HELP NOW MORE THAN EVER. THIS ONCE-BOUNTIFUL RAINFOREST IS DOWN TO JUST 2% OF ITS FORMER COVERAGE, WITH SEVERE WATER CONTAMINATION LEVELS ON THE RISE. YOUR PURCHASE OF THIS BOOK WILL DIRECTLY BENEFIT RESERVA'S CONSERVATION EFFORTS.

FIN

Barb and Rudy

WRITTEN AND DRAWN BY
THE PERRY BIBLE FELLOWSHIP

LOSS OF OUR LAND

WRITTEN AND DRAWN BY
ARIELA KRISTANTINA
CONSULTING EXPERT
BRADLEY HILLER

"THE FATE OF INDONESIA'S FORESTS AND PEATLANDS COULD MAKE OR BREAK THE WORLD'S GOAL TO LIMIT GLOBAL WARMING..."

– SIR NICHOLAS STERN.

WE ALL MAY BE LEAVING SOON.
DROUGHT GETS LONGER. IT'S HARD TO GET FRESH WATER.
THE HEAT AND SMOKE CHOKE MY WHOLE FAMILY.
I AM SO SORRY THIS IS HAPPENING. WE SHOULD DO BETTER.
I AM SO SORRY.

THE UNITED NATIONS IDENTIFIES TROPICAL PEATLANDS ACROSS SOUTHEAST ASIA, AFRICA, THE CARIBBEAN, AND SOUTH AND CENTRAL AMERICA. THESE ECOSYSTEMS PROVIDE IMPORTANT CARBON SEQUESTRATION AND WATER REGULATION SERVICES, AS WELL AS SUPPORTING LOCAL LIVELIHOODS AND RICH BIODIVERSITY.

HOWEVER, TROPICAL PEATLANDS ARE INCREASINGLY VULNERABLE TO HUMAN- AND CLIMATE-INDUCED THREATS, INCLUDING DEFORESTATION, DRAINAGE, AND FIRE. GIVEN THAT MANY OF US GLOBALLY CONSUME PRODUCTS WHICH – DIRECTLY OR INDIRECTLY PLACE PRESSURES ON THESE SYSTEMS – FOR EXAMPLE, HALF OF PACKAGED PRODUCTS IN SUPERMARKETS TODAY CAN CONTAIN PALM OIL – WE HAVE A SHARED RESPONSIBILITY FOR THEIR SUSTAINABLE MANAGEMENT.

WHEN THESE FORESTS AND PEATLANDS ARE GONE, THE REST OF THE ECOSYSTEM WILL SOON FOLLOW. US INCLUDED.

FOREST HERO
INSPIRED BY
PAULO ADARIO
WRITTEN BY
DESIREE ALVARADO
ART BY
DAVID BAILLIE

WTK: FORA DA TERRA DENI!
GREENPEACE
WTK: OUT OF DENI LAND!
GREENPEACE
GREENPEACE

KAMARANPI...
THE UNIVERSAL LANGUAGE
DEVELOPED WITH
KT TUNSTALL
WRITTEN BY
KT TUNSTALL & TYLER JENNES
ART & COLORS BY
CRAIG CERMAK
LETTERING BY
BERNARDO BRICE
YOU'LL HEAL OVER SOME DAY...
PERU
HER FACE IS A MAP OF THE WORLD, IS A MAP OF THE WORLD...
I FEEEEEEL IT ALL...
YOU CAN SEE SHE'S A BEAUTIFUL GIRL, SHE'S A BEAUTIFUL GIRL...

WITH SPECIAL THANKS TO
ADAM GARDNER, JAMES VALENTINE, STEFAN LESSARD, DIANA RÍOS, & NICOLÁS SABA

SPECIES got TALENT
TONIGHT... WILL BE A MEMORY TOO...
AND A NEW DAY HAS BEGUN...

CLAP!
CLAP!
CLAP!

UP NEXT.. DRACULA ORCHID!

HI!

I'D REALLY LOVE TO BE SAVED FROM EXTINCTION. I THINK I HAVE A LOT TO OFFER...

I JUST WANT TO GET THE CHANCE TO PROVE MYSELF, JUST LIKE EVERY OTHER ORCHID.

DISQUALIFIED!

WELL... WHAT DID THEY SAY?
SORRY, EVERYONE...
ANIMALS ONLY...
IT'S OKAY, DRAC...
WE'LL GET THEM NEXT TIME.
MAY THE BEST ENDANGERED SPECIES WIN!
DEVELOPED WITH CALLIE BROADDUS
WRITTEN BY TYLER JENNES
ART AND LETTERING BY SHEILA PUTRI
@SHEILASPLAYGROUND

GRANDMA'S AMAZING ADVENTURES

DEVELOPED WITH MALAIKA VAZ | WRITTEN BY PAUL GOODENOUGH | ART & LETTERING BY MOMO & POPO

JUST SHOOT US

WRITTEN BY PAUL GOODENOUGH | ART & LETTERING BY KEN CATALINO

BULLFIGHT

WRITTEN BY RICKY GERVAIS | ART & LETTERING BY ROB STEEN

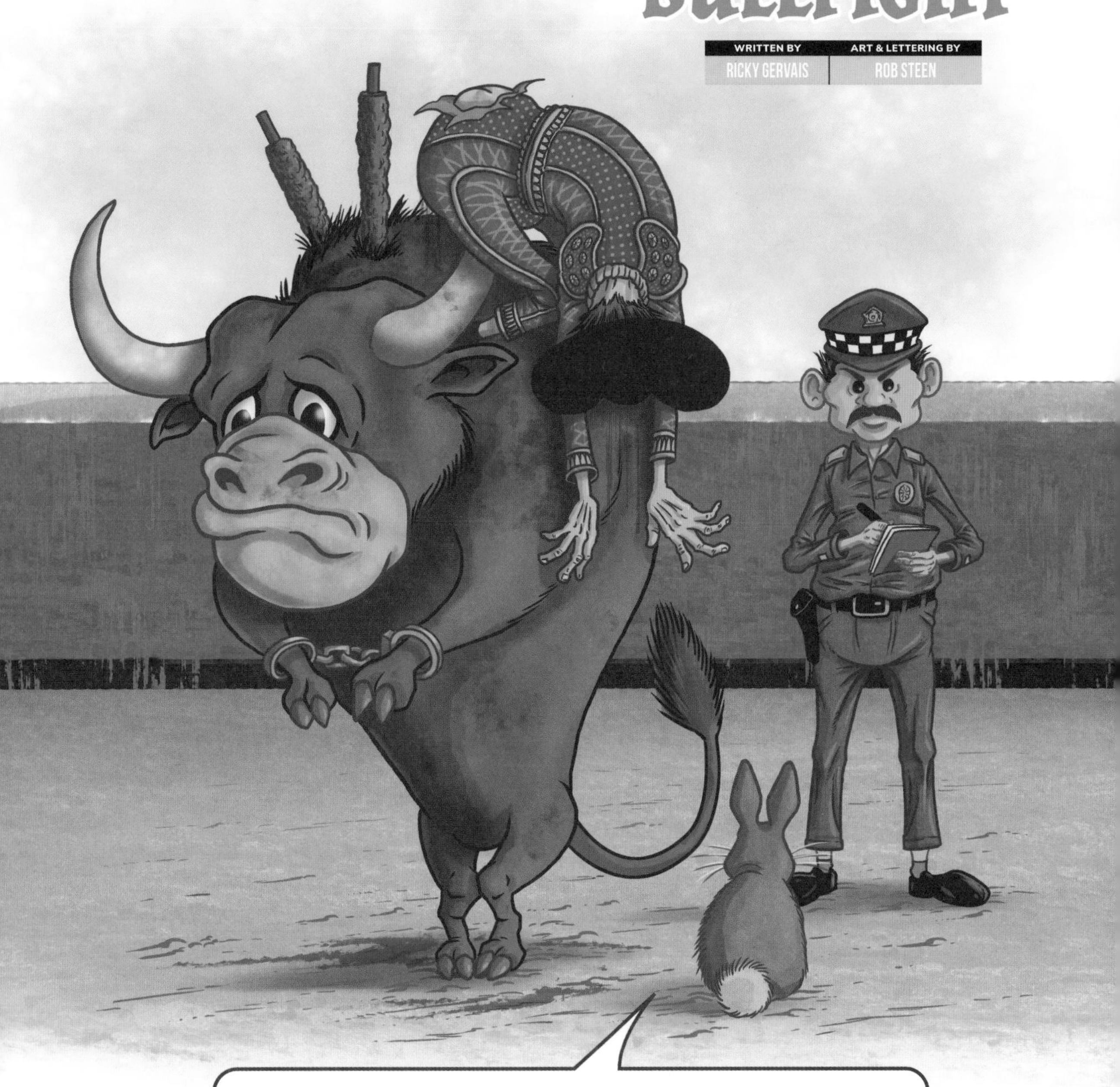

I SAW THE ***WHOLE THING***, OFFICER! THE MAN IN SPARKLY TIGHTS STARTED ***ATTACKING*** THE BULL FOR ***NO REASON***. THE BULL WAS JUST ***DEFENDING*** HIMSELF!

There's No One Like Me

WRITTEN AND DRAWN BY
SAD ANIMAL FACTS

sad animal facts

Hey Mommy!
What is that thing on your belly?
Oh, that...

Uhm, well...
And that thing Poking out?

And will they do that to me too?

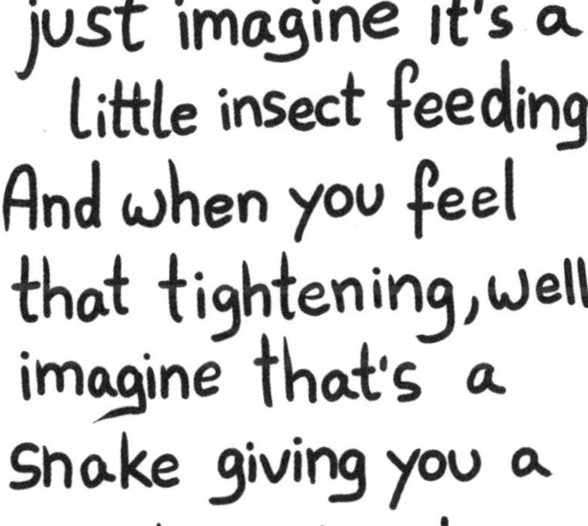
Look, darling... remember when we lived free in the big forest? Well, when you feel that jab, just imagine it's a little insect feeding. And when you feel that tightening, well imagine that's a snake giving you a biiiiig hug!

Woah!
But does it hurt?

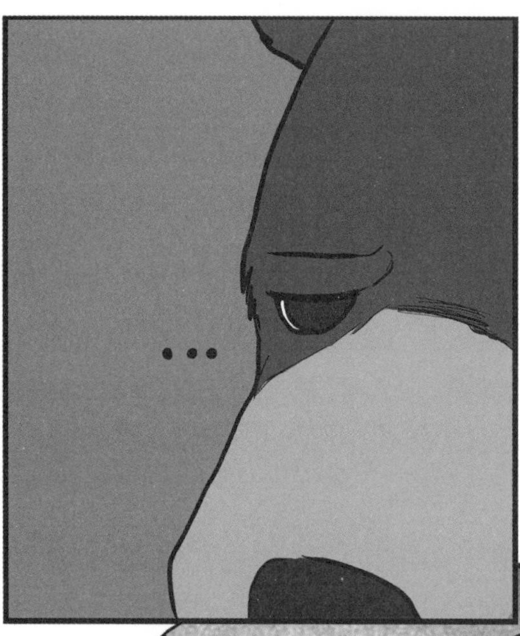
...

Hey, you know what?

If they move me again, or if ever you can't feel me near you...

... then just sing
"Mama bear Mama bear When you're here I'm not scared!"
And then I hear you and sing back to you!

I miss hugging you, mommy...

Mom!
Oh oh!

NOOO!
Don't be scared!
I'll always be with you!

Oh ...
Look at this cruelty
Mama bear mama bear
Hm? Do I hear singing from this shed?

Mama Bear

DEVELOPED WITH	WRITTEN BY	ART & LETTERING BY
WORLD ANIMAL PROTECTION	JENNY JINYA & PAUL GOODENOUGH	JENNY JINYA

Thousands of bears are suffering in the bear bile industry. Forced to endure a lifetime in captivity, they are exploited for their bile — an ingredient used in some traditional medicines that is painfully extracted from the bears' gallbladder.

World Animal Protection is working to end bear bile farming. Together with other NGOs, local partners, and governments, they aim to change consumer behaviour, improve legislation and law enforcement, promote better welfare for captive bears, and strive to ensure no new bears enter the industry.

For more information visit:

www.worldanimalprotection.org.uk

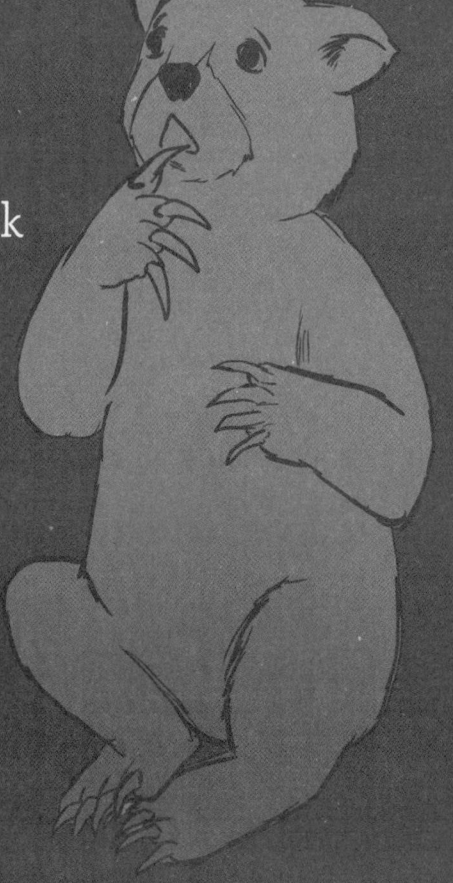

EARTH DAY

WRITTEN AND DRAWN BY
DAMI LEE

Sanctuary
Ellie's story
DEVELOPED WITH
PROF. BEN GARROD AND JIM & JENNY DESMOND
WRITTEN BY
DAVID BAILLIE
ART & COLORS BY
CONOR DOYLE
LETTERING BY
PAUL GOODENOUGH
KRAK KRAKAK
WAAAW! WAW!

WAAA-

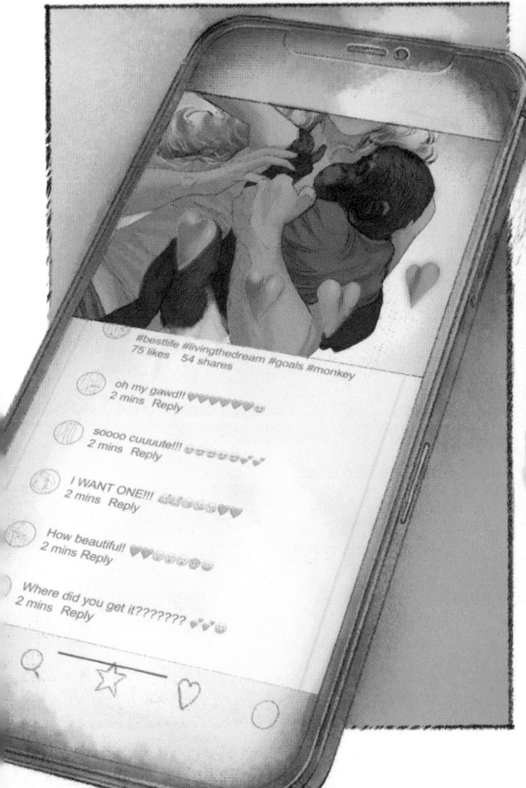
#bestlife #livingthedream #goals #monkey
75 likes 54 shares
oh my gawd!!
2 mins Reply
soooo cuuuute!!!
2 mins Reply
I WANT ONE!!!
2 mins Reply
How beautiful!
2 mins Reply
Where did you get it???????
2 mins Reply

-OOH?

WAAAW!
WAW!

FORESTRY

Liberia Chimpanzee Rescue & Protection is the first and **only** chimpanzee sanctuary and conservation center in Liberia.
LIBERIA CHIMPANZEE RESCUE & PROTECTION

Here they rescue and provide lifetime care to chimpanzees who are victims of the illegal bushmeat and pet trades.

And this is Ellie now, and this is her story of finding sanctuary.

Wise Tree

WRITTEN AND DRAWN BY
WAR AND PEAS

War and Peas

WRITTEN AND DRAWN BY
DINOS AND COMICS

dinos and comics

BRIDGE TO THE FUTURE

WRITTEN BY
RUTH FLETCHER GAGE & CHRISTOS GAGE
ART BY
MIKE PERKINS
LETTERING BY
BERNARDO BRICE

His name is **P-22**. He rules the Hollywood Hills. And he's the most famous mountain lion in the world.

HOLLYWOOD

He has his own social media accounts, merchandise, even plush toys.

But like too many true Hollywood stories, there's a tragedy underneath the glitz and glamour.

P-22 is alone.

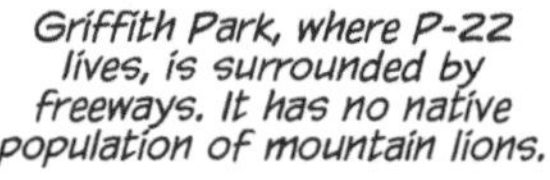

Griffith Park, where P-22 lives, is surrounded by freeways. It has no native population of mountain lions.

Surrounding mountain ranges do. And when cougar cubs grow up, their instinct is to find their own territory. But when they try to cross a freeway...

...it doesn't end well.

P-22 crossed **two** freeways. He's famous because he survived-- something no other big cat had. He's got prey, territory, and fame.

But he has no one to share it with.

But there's a way to **change** that. They're called **Wildlife Corridors.**
Bridges or tunnels designed for animals to get past freeways without risking being hit by cars.
They've already succeeded in other parts of the country and the world.
A bridge in Utah went viral thanks to a YouTube video of all the animals using it.
India and Africa have built massive elephant corridors.
The short-term benefit is the safety of both animals and humans. The long-term benefit...
...is **love.**

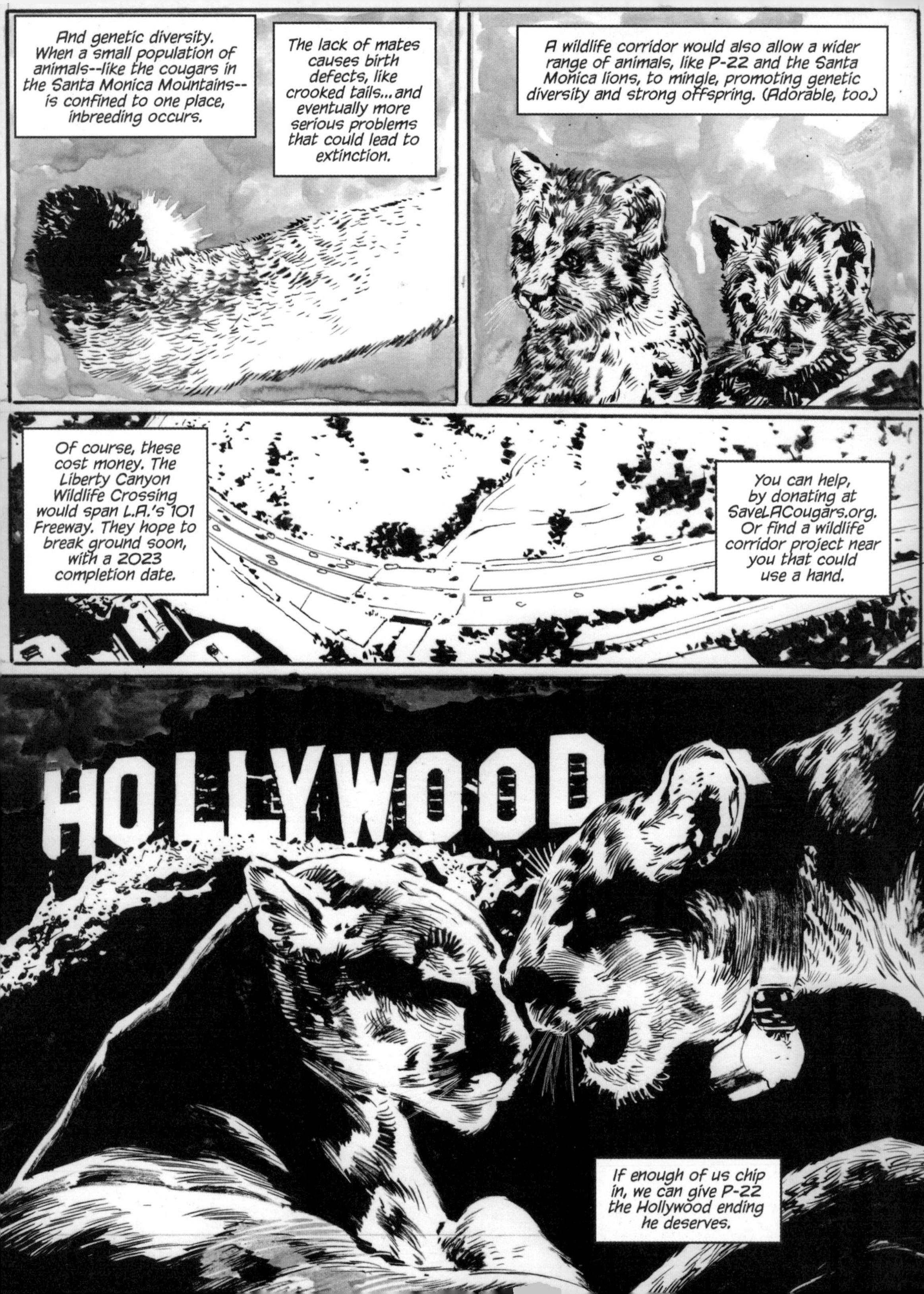

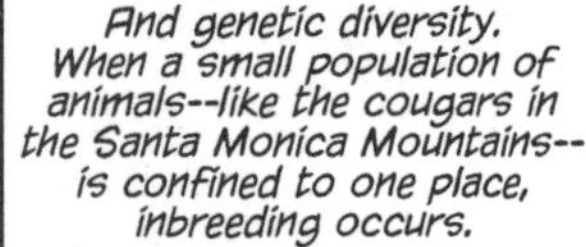

And genetic diversity. When a small population of animals--like the cougars in the Santa Monica Mountains--is confined to one place, inbreeding occurs.

The lack of mates causes birth defects, like crooked tails...and eventually more serious problems that could lead to extinction.

A wildlife corridor would also allow a wider range of animals, like P-22 and the Santa Monica lions, to mingle, promoting genetic diversity and strong offspring. (Adorable, too.)

Of course, these cost money. The Liberty Canyon Wildlife Crossing would span L.A.'s 101 Freeway. They hope to break ground soon, with a 2023 completion date.

You can help, by donating at SaveLACougars.org. Or find a wildlife corridor project near you that could use a hand.

HOLLYWOOD

If enough of us chip in, we can give P-22 the Hollywood ending he deserves.

BORN FREE FOREVER

DEVELOPED WITH	WRITTEN BY	ART & COLORS BY	LETTERING BY
VIRGINIA MCKENNA OBE	TAB MURPHY	MWELWA TAX	JIM CAMPBELL

FATHER? IS EVERYTHING OKAY?

I WAS JUST THINKING ABOUT YOUR GREAT-GREAT-GREAT-GRANDMOTHER.

BACK IN HER DAY, THINGS WERE VERY DIFFERENT.
AFRICA SEEMED ENDLESS, OUR NUMBERS INFINITE.
SIGH
THESE DAYS...WE ARE JUST TRYING TO HANG ON.
WHAT HAPPENED, FATHER? WHY ARE WE SO FEW?
MANY REASONS, SON.

"WE'VE BEEN PERSECUTED AND HUNTED...

"OUR HABITAT LOST...

"OUR NATURAL PREY DEPLETED...

HUMANS...
DO THEY HATE US?
NO, SON, THEY DON'T HATE US.
BUT SOMETIMES THEY ONLY THINK ABOUT THEMSELVES. WHEN WE'RE GONE, IT WILL BE TOO LATE.
DON'T BE SCARED. THERE ARE GOOD PEOPLE OUT THERE. FIGHTING FOR US. MAKING A DIFFERENCE. WE *WILL* CLAW OUR WAY BACK.
AFTER ALL, WE'RE THE KING OF THE BEASTS.

Dedicated to Elsa the Lioness.

'Who will care for the animals for they cannot look after themselves? Are there young men and women willing to take on this charge? Who will raise their voices to plead their case when mine is carried away on the wind?'

George Adamson 1906 - 1989

Hedgehog Heroes!

INSPIRED BY MICHAEL ROSEN

WRITTEN BY HUGH WARWICK

ART & LETTERING BY STIG

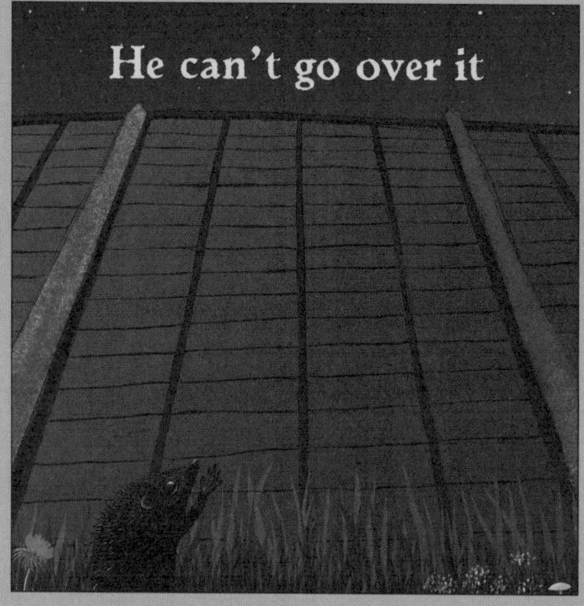
He can't go over it

He can't go under it

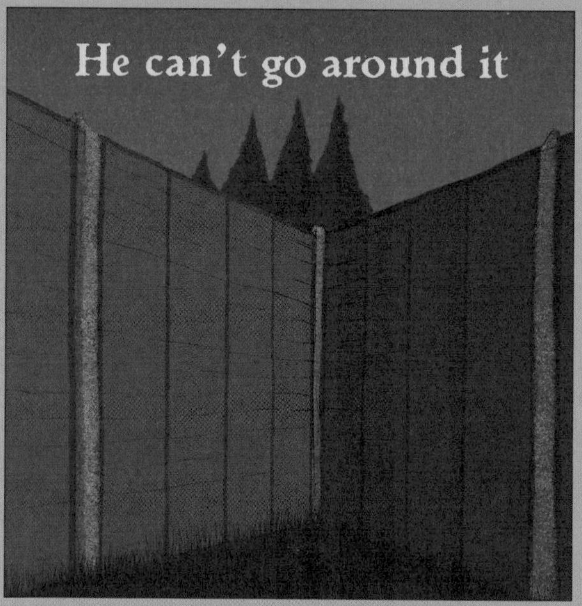
He can't go around it

But she got through it!

Help hedgehogs find love - Make holes in your fences!

Create a Hedgehog Street

What if God was one of us...

INSPIRED AND LYRICS BY	WRITTEN AND DRAWN BY
ERIC BAZILIAN	ALICIA SOUZA

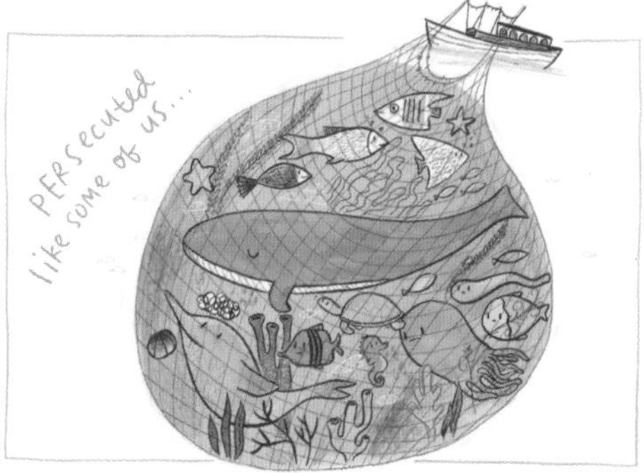

SHARED EXPERIENCES

STORY INSPIRED BY	WRITTEN AND DRAWN BY
PAUL, MARY, AND STELLA MCCARTNEY AND MEAT FREE MONDAY	JUST COMICS

Never Forget

DEVELOPED WITH	WRITTEN BY	ART & LETTERING BY
BROOKLYN BECKHAM AND WORLD ANIMAL PROTECTION	PAUL GOODENOUGH	BUDDY GATOR

END WILDLIFE CRIME

BY JOHN SCANLON. ILLUSTRATION BY MISTER HOPE.

Hundreds of thousands of wild animals and plants are trafficked every year, from pangolins to elephants, orchids to rosewood, eels to sharks.

During the last decade, more than **a million pangolins** were trafficked, with more than **128 tons** of their scales and meat seized by the authorities in just one year.

An astonishing **1.7 million** undiscovered viruses are thought to exist in animals living in the wild, half of which could spill over to people, and this could be devastating for our health—yet, the numbers of trafficked wild animals keep rising at alarming levels.

Wildlife trafficking drives species toward extinction, decimates ecosystems and their ability to sequester carbon, results in governments being deprived of revenue, and leads to more corruption, insecurity and poverty.

The impact of these crimes on ecosystems adds up to a gar gantuan **$1–2 trillion** a year in value.

HELP KEEP WILDLIFE IN THE WILD!

1. Be careful what you buy, and never buy animals as pets that have been taken from the wild. Let wildlife remain in its natural habitat.

2. Share this message—tell your friends and family about the devastating destruction that wildlife trafficking is causing.

3. Voice the importance of scaling up wildlife law enforcement to politicians and the world so we can end wildlife crime.

£250,000
Tackle pangolin trafficking in Eastern India, and fight the loss of biodiversity caused by the international wildlife trade.

£50,000
African Lions: Protect community livestock and prevent predator revenge attacks in Southern Kenya.

£25,000
Eastern Lowland Gorilla: Collaborate with the Pole Pole Foundation in DRC, and protect elephants and giraffes in Northern Kenya.

£10,000
Raise the Red Flag: Expose the negative human exploitation of all wild animals globally.

THE MONEY WE RAISE IS HELPING:

KEEP WILDLIFE IN THE WILD

With nearly four decades of experience, Born Free cares for and conserves some of the world's most iconic, yet persecuted, wild animal species, safeguards habitats, empowers local communities, and inspires current and future generations to look after all life on Earth and keep wildlife in the wild.

WHY IS THE PROJECT SO IMPORTANT?

Every project Born Free undertakes is vitally important. Exposing the brutal impact of the international wildlife trade on endangered species; protecting Eastern lowland gorillas and their habitat; helping communities safeguard their livestock and, thereby, preventing the killing of threatened lions; working with communities to conserve wild tigers; spearheading landscape-level efforts to protect ecosystems, such as the Dja Biosphere Reserve in Cameroon and the iconic species that live there. Our unique approach, which we call compassionate conservation, places the welfare of individual wild animals at the heart of our strategies and actions. Our milestones are ambitious, practical, and transformative.

Each milestone we reach, each successfully funded project, will enable Born Free to increase our support for key wildlife species; enhance our engagement with local communities; apply effective, innovative solutions; and tackle the many pressures facing wildlife and our natural world.

SPECIES PROFILES

Wild orangutans are threatened by habitat loss, illegal hunting, and trade in infants as "pets." Just 20,000 African lions remain in the wild today, down 80% in fifty years. In the last decade, more than one million wild pangolins have been poached for meat and scales.

> ***The natural world is vulnerable, as never before. It's time for compassionate, caring individuals to come together and protect all that we hold dear.***

Virginia McKenna, OBE co-founder. The Born Free Foundation

AMBOSELI, KENYA

One elephant is killed for its tusks every twenty minutes.

PROJECT IN ACTION

Born Free has built more than 330 "predator-proof bomas" that protect the livestock of pastoral communities at night. If their cattle, sheep, and goats are safe, local people tolerate lions and other predators. So far we have helped more than 6,000 people, 80,000 head of livestock–and the wild predators of southern Kenya.

THE FUTURE

Every project we undertake helps Born Free secure a better future for iconic species–lions, tigers, elephant, rhino, gorillas, chimpanzees, orangutans, pangolins, giraffes, and more. We always treat individual wild animals with respect and compassion, as part of functional, viable ecosystems.

FIND OUT MORE

> SIGN UP AT WWW.BORNFREE.ORG.UK TO GET THE LATEST NEWS ABOUT OUR WORK AND HOW YOU CAN HELP DELIVERED STRAIGHT TO YOUR INBOX.

> FOLLOW BORN FREE FOUNDATION ON SOCIAL MEDIA FOR CONTINUOUS UPDATES ON OUR AWARD-WINNING CONSERVATION WORK.

Even though
I feed you
as much as
I can...

... you're never
full.

This is all I can find

You get weaker every day

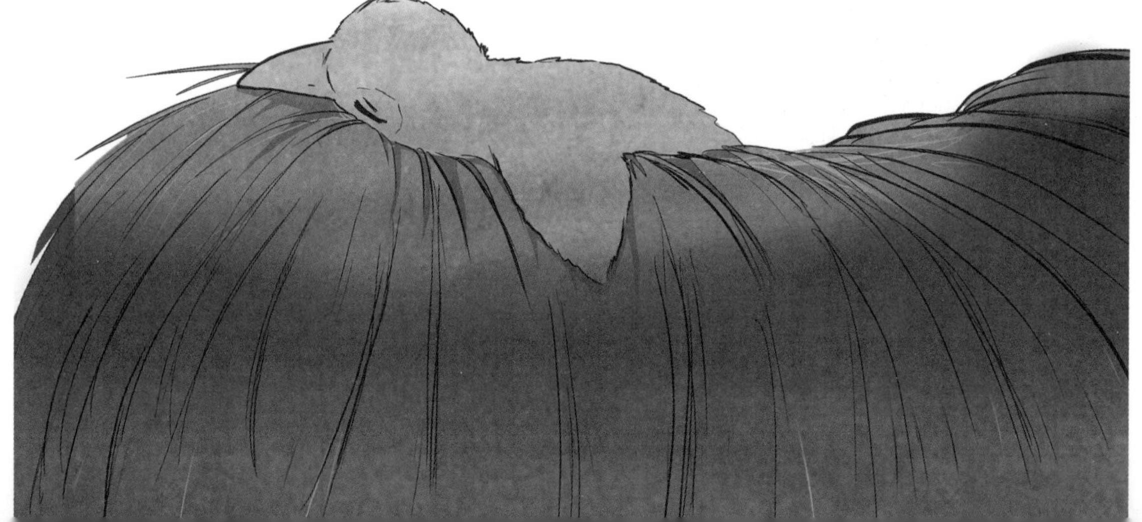

But, you know …

…I'm getting weaker too.

My stomach hurts

you haven't failed your chick
you did everything you could
Come here, little one.

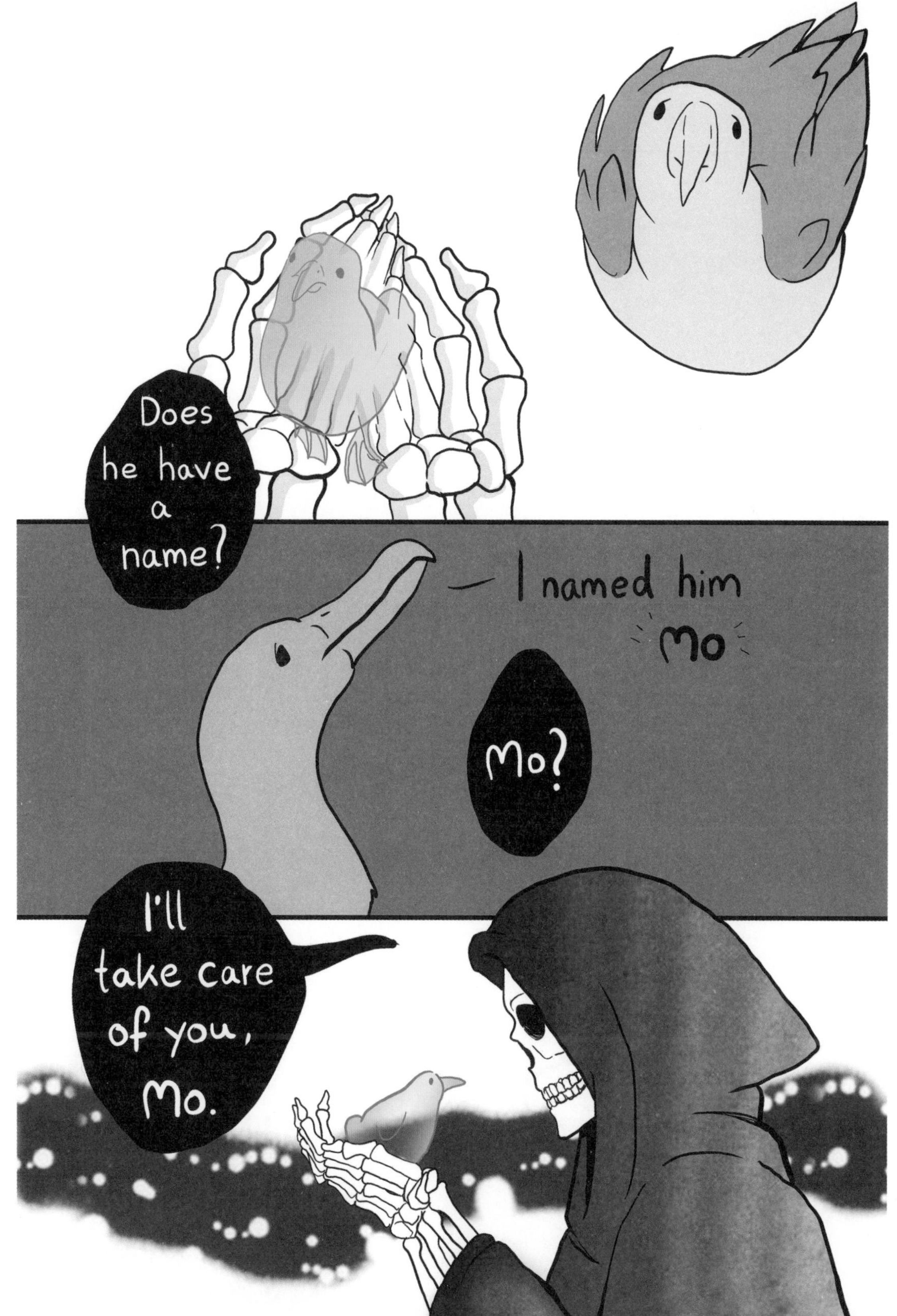
Does he have a name?
I named him
Mo
Mo?
I'll take care of you, Mo.

Here
Say good-bye
Forgive me, Mo
farewell...
I love you

Plastic trash is found in 90 percent of seabirds, revealed in a new study*.

"Plastic found inside birds includes bags, bottle caps, synthetic fibers from clothing, and tiny rice-sized bits that have been broken down by the sun and waves. The study found a 67 percent decline in seabird populations between 1950 and 2010. Albatross are more prone to eating plastic because they fish by skimming their beaks across the top of the water, and inadvertently take in plastics floating on the surface."

Sources:

https://www.nationalgeographic.com/news/2015/09/15092-plastic-seabirds-albatross-australia/

*https://www.pnas.org/content/early/2015/08/27/1502108112

Little Mo

WRITTEN AND DRAWN BY
JENNY JINYA

WRITTEN AND DRAWN BY
BUDDY GATOR

I had a nightmare last night, it was so scary.
What did you saw in your nightmare? A ghost? A monster?

A straw.

TORN

DEVELOPED WITH
KIDS AGAINST PLASTIC
ART AND WRITTEN BY
AMY MEEK
CLIMATE IN CRISIS
12 YEARS LEFT!

GOVERNMENT ACTION IN NEW 2025 GREEN DEAL
1.5 REACHED ... WHAT NEXT?!
WE NEED TO ACT!
GOVERNMENT INACTION

2030 TARGETS MET
MISSED DEADLINES
IT'S 2030 & TOO LATE

CLIMATE RECOVERY
CRISIS CONTINUES

IT STARTED WITH A PICTURE.
IT SHOOK ME TO MY CORE.
THE MASSIVE WASTE MANAGEMENT PROBLEM IN MY HOME, INDIA.
I HAD TO TAKE THE PROBLEM ON. AT THE SOURCE.
THE POWER OF YOUTH
DEVELOPED WITH
ADITYA MUKARJI
WRITTEN BY
ALEX SEGURA
ART & COLORS BY
GEORGE KAMBADAIS
LETTERING BY
JIM CAMPBELL
I WAS A CHILD, THOUGH. WHAT CAN A CHILD REALLY DO?
A LOT, I LEARNED.

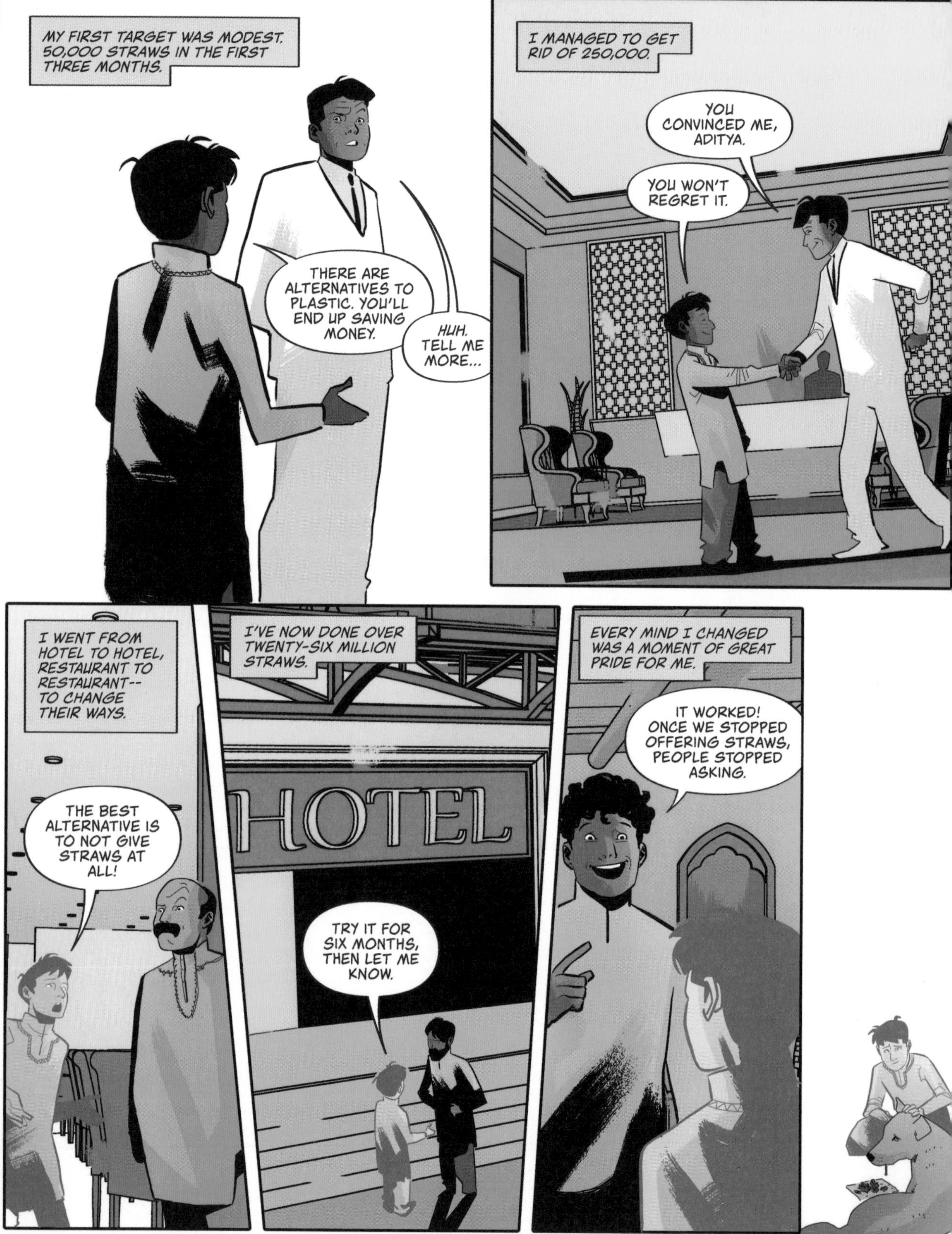

MY FIRST TARGET WAS MODEST. 50,000 STRAWS IN THE FIRST THREE MONTHS.
THERE ARE ALTERNATIVES TO PLASTIC. YOU'LL END UP SAVING MONEY.
HUH. TELL ME MORE...
I MANAGED TO GET RID OF 250,000.
YOU CONVINCED ME, ADITYA.
YOU WON'T REGRET IT.
I WENT FROM HOTEL TO HOTEL, RESTAURANT TO RESTAURANT-- TO CHANGE THEIR WAYS.
THE BEST ALTERNATIVE IS TO NOT GIVE STRAWS AT ALL!
I'VE NOW DONE OVER TWENTY-SIX MILLION STRAWS.
HOTEL
TRY IT FOR SIX MONTHS, THEN LET ME KNOW.
EVERY MIND I CHANGED WAS A MOMENT OF GREAT PRIDE FOR ME.
IT WORKED! ONCE WE STOPPED OFFERING STRAWS, PEOPLE STOPPED ASKING.

THEN I MOVED INTO RAISING PUBLIC AWARENESS. I HELPED OTHERS LEARN WHAT THEY COULD DO.
WHAT, SO NO STRAWS?
JUST GIVE IT A TRY. TRUST ME.
THE YOUTH ARE INTEGRAL. THEY HAVE MORE EMOTION.
THEY DELIVER THE MESSAGE WITH PASSION AND HONESTY.

WHY EVEN GIVE OUT STRAWS? IT'S HURTING OUR PLANET.
I'M JUST TRYING TO DO MY--
LET ME TALK TO THE BOSS, OKAY? THIS IS IMPORTANT.
AND ADULTS LISTEN.

REFUSE IF YOU CANNOT REUSE. LEAVE THE WORLD A LITTLE BETTER THAN YOU FOUND IT.

NEVER BE AFRAID TO CHALLENGE THE STATUS QUO.

ALWAYS AIM HIGHER THAN YOUR TARGET.

KIDS ARE REFUSING SINGLE-USE PLASTIC, MORE AND MORE.
IT'S HAPPENING **NOW**. ACROSS THE PLANET.
AND THE MOVEMENT WILL ONLY **GROW**.
THIS STORY IS DEVOTED TO THE WORLD-CHANGING WORK OF ADITYA MUKARJI AND ALL THE YOUNG PEOPLE WHO HAVE SHOWN US THE WAY FORWARD.

million

Around the planet, children have been responsible for removing more than 30 million pieces of plastic from the environment.

#savetheplanet

POSTER DESIGNED BY
MOKSHA CARAMBIAH

PROJECT MILESTONES

£1,000,000
Would allow FUNDAECO to purchase and protect 1,788 acres (723.5 hectares) of the Laguna Grande Reserve. The protected area will provide habitat for the jaguar.

£100,000
Purchase and protect 178 acres (72 hectares) of the Laguna Grande Reserve. The protected area will provide habitat for the Baird's tapir.

£50,000
Purchase and protect 89 acres (36 hectares) of the Laguna Grande Reserve. The protected area will provide habitat for the Yucatan black howler monkey.

£10,000
Purchase and protect 17 acres (6.8 hectares) of the Laguna Grande Reserve. The protected area will provide habitat for the margay.

THE MONEY WE RAISE IS HELPING:

GUATEMALA'S LAGUNA GRANDE RESERVE

Acting as a natural water filter, flood barrier, and carbon sink, mangroves play a crucial role in ecosystems the world over. In Caribbean Guatemala, the 316,000-acre Laguna Grande project does all this and more, tackling the climate and biodiversity crises while providing sustainable livelihoods for local Indigenous communities.

WHY IS THE PROJECT SO IMPORTANT?

Landscapes like Laguna Grande bring benefits to conservation, communities, and climate, but without the proper protection these vital services could be lost forever. Supporters of Laguna Grande can help to preserve this vast and vitally important habitat, preventing the draining and deforestation that comes with unchecked development.

This is an ecosystem that captures and stores huge amounts of carbon, helping us win the fight against climate change. It also harbors iconic species like the jaguar, Baird's tapir, and West Indian manatee, and defends one of the world's most climate-vulnerable nations from natural disasters. Meanwhile, project partner FUNDAECO work tirelessly to support sustainable livelihoods and provide health services to communities.

This project will support the expansion of the Laguna Grande Reserve, protecting an additional 3,294 acres (1,333 hectares) of forest. Threatened by timber extraction, the forest is part of the Mesoamerican Biological Corridor, home to more than 350 bird species.

SPECIES PROFILE

Maintaining connectivity between habitats is critical for far-ranging predators like the jaguar, and this is exactly what purchasing land in Laguna Grande will achieve. Bringing more land under protection will allow rangers to expand their patrols, keeping the Americas' iconic big cat safe from poachers.

> "The money that is given to the World Land Trust, in my estimation, has more effect on the wild world than almost anything I can think of."

Sir David Attenborough, World Land Trust patron

LAGUNA GRANDE, GUATEMALA

A mosaic of lagoons, mangroves, reefs, and tropical forests, Laguna Grande is a prime example of a thriving wetland ecosystem.

THE FUTURE

By protecting much-needed habitat for threatened species and providing a natural solution to the climate crisis, the success of Laguna Grande will help change the world for the better. With benefits shared between people and wildlife, the project can also offer a blueprint for future conservation initiatives in Guatemala and beyond.

PROJECT IN ACTION

Project funds are utilized on the ground by FUNDAECO, a conservation organization that has worked in the area for more than twenty years. The management of Laguna Grande is conducted in partnership with local Indigenous association Amantes de la Tierra.

FIND OUT MORE

> VISIT: WWW.WORLDLANDTRUST.ORG/WHAT-WE-DO/CARBON-BALANCED
> VISIT: WWW.WORLDLANDTRUST.ORG/APPEALS/ACTION-FUND

We're working with five charities to undertake specific, targeted projects across the globe to undo some of the catastrophic damage done to the planet.

From rewilding, to species reintroduction and restoration, our project will put species back into recovery, enabling them to look after themselves, forever.

Sponsored projects:

RESERVA
THE YOUTH LAND TRUST

EXPLORING MY OPTIONS

you're the only one for me, babe

This orchid is called *paphiopedilum superbiens.*
Do you see how it grows straight up from the floor?
It got its name because it looks like a slipper!
And this one's a bulbophyllum lobbii!
Do you smell that?
It produces a smell like rotten meat to attract flies and eat them!
The Smallest Seed
DEVELOPED WITH
PUNGKY NANDA PRATAMA
WRITTEN BY
SARAH FLORENCE LORD
ART & COLORS BY
ABBY HOWARD
LETTERING BY
BERNARDO BRICE
And this one is *taeniophyllum obtusum.*
Some people call it the "ghost orchid" because it is so small.
It only produces one or two flowers, less than a millimeter in diameter each!
Orchids are very fragile beings.
They have microhabitats, which means they need very specific elements. Lots of these plants are protected by the government, but it doesn't stop bad people from hunting and selling them.
That's why *we* have to look after them.
When a habitat is destroyed, it's a disaster.
An animal can relocate-- a plant can't!

These plants will be lost forever if people don't stop destroying their environment.
And these plants bring life to us all!
Our dream is to save every species of orchid.
So far we have over 400, but there are so many more...
I'm so, *so* happy that you're telling my story.
I'm sorry, I'm getting emotional.
I just want so badly everyone to hear this--so they'll care about plants like they do for animals.
It makes me so happy talking to you, and to everyone out there, because if people care, I know we can save these amazing species!
Thank you.
Thank you so much.

YELLOW BIRD
WRITTEN BY
TATE BROMBAL
ART & COLORS BY
JEFF LEMIRE & RAY FAWKES
LETTERING BY
BERNARDO BRICE
Rodrigues, Mauritius.
The Rodrigues Fody is a resilient bird.
This resilience comes from their intricately woven nests, often built several to a branch in clusters of colonies.
These communities of birds protect each other, doing their part to survive.
But resilience can only afford a bird so much.
Not when cyclones and invasive species and hungry livestock and greedy humans threaten their home.
Not when the trees and their colonies dwindle one by one...
Once abundant, by 1968, there were only five pairs of Rodrigues Fodies left.
They seemed a lost cause, another bird doomed to extinction-- not unlike its Mauritian cousin, the Dodo.
But then, something took flight...

Mauritius, and its people, took charge.
And an extensive conservation effort was put into place.
Deforestation was halted.
Trees were planted.
Invasive species were *uprooted.*
Boundaries were erected to give the forests, the animals, the Fodies their space to thrive.
And the weather calmed down.
And the roots held strong.
And the forests bounced back.
The Rodrigues Fody took flight once more.

Our home's own conservation effort has been mid-flight for some time now.
Held up somewhere between those unwilling to act and those deeming us a lost cause.
But nothing is lost while we still have each other.
RDEN
All it takes is the courage to fly. And even if your actions seem too small...
SALVE EL AMAZONIA
REJK FÖR KLIMATET!
SAVE THE CONGO
Or your voice too quiet...
Just know that we are strongest as one, protecting each other, doing our part to survive.
And if climate change ever seems too big...
Or if the road forward ever appears too insurmountable...
WATE IS LIFE
Just think of a small yellow bird once doomed in the handfuls, now soaring in flocks over 17,000.
Because, just like the Rodrigues Fody, we are all resilient...
...but we are only as resilient as our home.

Rewild the Galapagos

In the Galapagos Islands, people are intertwined with the wildlife and natural world surrounding them. Finches will land on your hand, while iguanas scurry past your feet, but not everything is perfect in paradise.

Local communities have seen animals they cherish disappear before their eyes, and more are teetering on the brink of extinction. Luckily, there is still time to reverse course.

We can and will rewild the Galapagos. It has become my life's mission to restore islands and reintroduce native animals, working alongside local communities to create ecosystems where people and wildlife thrive together.

The more people who know and support us, the more species we can save from extinction.

With love and hope. Paula A. Castaño

Art by Ian Stopforth

£250,000
Will pay for a predator-free sanctuary for Tasmanian devils and other native wildlife of East Australia to restore the ecological functioning of native forest ecosystems.

£100,000
Will pay for the build of a first-of-its-kind conservation breeding center for unique and threatened species of the Annamite Mountains.

£50,000
Feed and care for twenty Tasmanian devils for an entire year in a conservation breeding program to bring the Tasmanian devils back to mainland Australia.

£10,000
Supports a community member to remove snares from Pu Mat National Park in Vietnam for three years.

re:wild

THE MONEY WE RAISE IS HELPING:

REWILD OUR WORLD!

The most effective solution to the triple threat of climate chaos, mass extinction, and emerging diseases is to protect and restore Earth's most irreplaceable places for biodiversity. We don't need to reinvent the planet, we just need to rewild it!

WHY IS THE PROJECT SO IMPORTANT?

This project will recover and restore critical wild areas, with a focus on two unique ecosystems critical to our global biodiversity.

In the forests of East Australia, damaged ecosystems have been ravaged by fires. We are embarking on a plan to restore lost ecosystems one species at a time. Starting with the Tasmanian devil, which we are working with our partners to bring to mainland Australia after a three-thousand-year absence. Devils allow native small mammals to recover by keeping invasive cats and foxes in check. By burying leaf litter as they forage, these small mammals help sustain cooler and less damaging fires.

In the rugged Annamite Mountains of Vietnam and Laos roams an animal that no living scientist has spotted in the wild. We are working to bring the saola back from the brink to resume its place among a wondrous assemblage of unique species through an innovative conservation breeding program to rewild one of the most biodiverse corners of our planet.

SPECIES PROFILES

The Tasmanian devil is an iconic ecosystem engineer instrumental in rebuilding Australia's lost ecosystems. The saola is an elusive myth-like animal and flagship for the Annamite Mountains, also known as the "Asian Unicorn." The critically endangered northern white-cheeked gibbon plays an important role dispersing seeds in Vietnam and Laos.

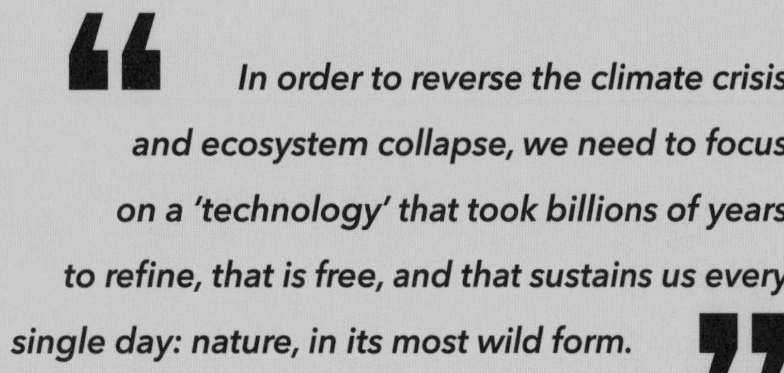

In order to reverse the climate crisis and ecosystem collapse, we need to focus on a 'technology' that took billions of years to refine, that is free, and that sustains us every single day: nature, in its most wild form.

Wes Sechrest, chief scientist and CEO of Re:wild

EASTERN FORESTS, AUSTRALIA

A global biodiversity hotspot that we are working to rewild one species at a time.

PROJECT IN ACTION

In Australia we work with local partners Aussie Ark to reintroduce Tasmanian devils into predator-free areas, and to monitor their success, building on years of a successful conservation breeding program. In the Annamites we work with WWF Vietnam, Bach Ma National Park, and Wroclaw Zoo. Communities around the project sites and conservation breeding centers benefit from direct employment and enhanced tourism potential.

THE FUTURE

Healthy, thriving ecosystems in Australia and the Annamites with a full complement of native wildlife will show that we can rewild some of the most biodiverse and irreplaceable places on Earth!

FIND OUT MORE

> FOLLOW REWILD ON SOCIAL MEDIA TO STAY UP TO DATE WITH OUR ONGOING CONSERVATION WORK.

THE CLOUD FOREST

DEVELOPED WITH
CALLIE BROADDUS
WRITTEN BY
MARGUERITE BENNETT
ART & COLORS BY
MAIA KOBABE
LETTERING BY
BERNARDO BRICE

THIS IS OUR HOME, THE BEAUTIFUL CHOCÓ CLOUD FOREST OF ECUADOR.
AND THE HUMANS MAKE IT SMALLER BY THE DAY.

THEY TAKE OUR TREES FOR CATTLE.
COME ON, I THINK THERE ARE A FEW MOUTHFULS OF GRASS OVER THERE--
HEY, I'M SLITHERIN' HERE!

THERE GOES THE NEIGHBORHOOD--
THEY TAKE OUR MOUNTAINS FOR GOLD.

WAITER, THERE'S SOUP ON MY FLY!
THEY TAKE OUR FOOD WITH PESTICIDES.

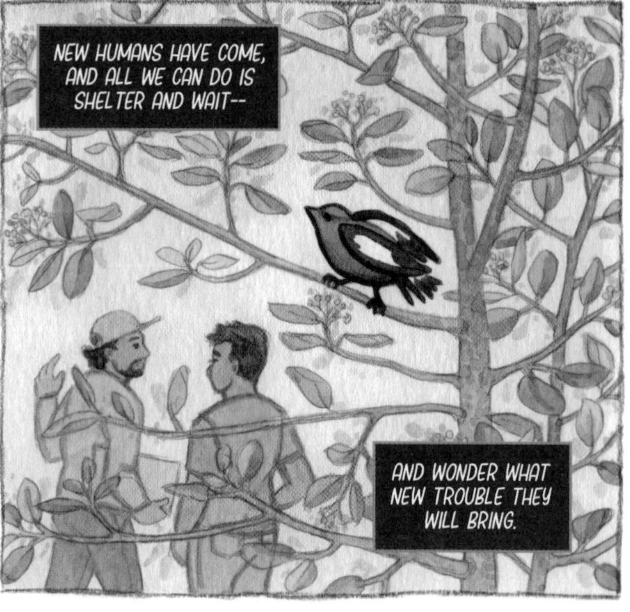

NEW HUMANS HAVE COME, AND ALL WE CAN DO IS SHELTER AND WAIT--
AND WONDER WHAT NEW TROUBLE THEY WILL BRING.

BUT INSTEAD...THEY MEAN US NO HARM.
THEY HAVE COME TO PROTECT US. THEY'VE COME TO LEARN FROM US.
THEY DISCOVER, DOCUMENT, AND LET US GO.

THEY SAY WE ARE RARE BEYOND RARE. THAT ONLY WITHIN THIS EVER-DWINDLING CLOUD FOREST ARE WE FOUND.

THAT SOME OF US WERE THOUGHT LOST, GONE FOREVER FROM THE EARTH, AND ARE REBORN, REDISCOVERED BY THESE HUMANS WHO COME TO HELP.

FOR SO LONG, WE HAVE BEEN HIDDEN--HAVE BEEN HIDING, AND SCARED, AND SILENT.
BUT NOW THAT YOU KNOW WE ARE HERE...

FINALLY, I HOPE WE CAN UNDERSTAND ONE ANOTHER.
THANK YOU! THANK YOU! THANK YOU...

£500,000
Purchase and protect 1,000 acres of habitat to secure a wildlife corridor for Ecuador's iconic Chocó species (puma, spectacled bear, ocelot, and more) and empower 231,660 young people.

£250,000
Purchase and protect 500 acres of habitat for brown-headed spider monkey and empower 115,830 young people.

£50,000
Purchase and protect 100 acres of habitat for the black-and-chestnut eagle and empower 23,166 young people.

£10,000
Purchase and protect twenty acres of habitat for the Rio Faisanes stubfoot toad and empower 11,583 young people through the 1 Million Letters #ForNature program.

RESERVA
THE YOUTH LAND TRUST

THE MONEY WE RAISE IS HELPING:

A FUTURE FOR THE CHOCÓ

The Chocó is one of the most ecologically unique places on Earth, but rampant deforestation for ranching, agriculture, and extractive industries like gold mining and commercial logging are quickly reducing this critically biodiverse zone to a fraction of its original size.

WHY IS THE PROJECT SO IMPORTANT?

While scientists race to study this area's endangered, new-to-science, and endemic species—that live here and nowhere else—conservation groups are working together to build a network of strategic protected areas. They are purchasing plots of intact habitat that might otherwise be sold for commercial development and employing local and Indigenous people as forest guardians.

After establishing an entirely youth-funded nature reserve here in 2021 in partnership with Rainforest Trust and Fundación EcoMinga, Reserva is again championing a creative youth-led effort to protect what remains of this irreplaceable forest by matching letters from youth with actual funds for conservation.

This funding will support the purchase and protection of Ecuadorian Chocó cloud forest, securing habitat for the critically endangered brown-headed spider monkey, black-and-chestnut eagle, iconic species like the puma and spectacled bear, and thousands of other species—empowering young people along the way.

SPECIES PROFILES

The intricately patterned Rio Faisanes stubfoot toad is one of the most endangered amphibians in Ecuador, relying on the expansion of Dracula Reserve for its survival. The Brown-headed spider monkey is one of the top twenty-five most endangered primates in the world and one of the three species of monkey on this site. The Chocó is home to more than fifty endemic bird species, like this multicolored plate-billed mountain toucan.

> *"The first step to stopping extinction is halting the loss of wild habitat. We have to stop the bleeding. Literally."*

Callie Broaddus, founder & executive director, Reserva: The Youth Land Trust

DRACULA YOUTH RESERVE, ECUADOR

The current edge of Dracula Youth Reserve is called Peñas Blancas (white rocks), situated in a conservation corridor established over eight years of international collaboration.

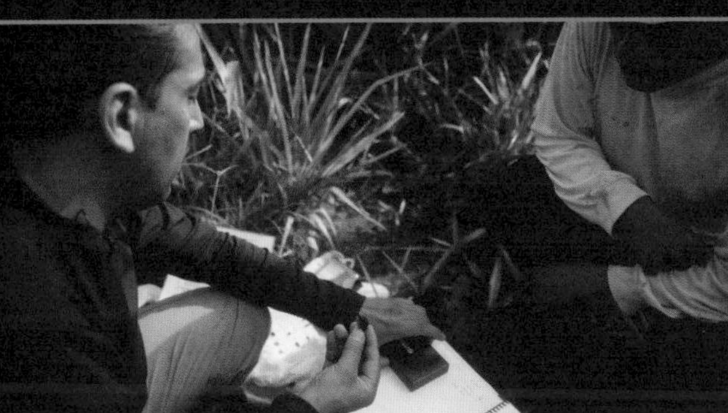

PROJECT IN ACTION

Funding for this project covers the cost of purchasing, protecting, and understanding this critically biodiverse landscape, including the expeditions necessary to monitor wildlife and salaries for local and Indigenous staff who are hired to protect these fragile ecosystems. Land will be owned and managed by Ecuadorian NGO Fundación EcoMinga.

THE FUTURE

The Ecuadorian Chocó is uniquely understudied, with nearly every expedition unearthing a new-to-science frog, mouse, or orchid. By empowering youth to have a hand in protecting this ecosystem both from afar and through local engagement, this project is also protecting the future of discovery for a generation of children and young scientists growing up with dwindling wild places to explore.

FIND OUT MORE

> CONTRIBUTE TO OUR 1 MILLION LETTERS #FORNATURE CAMPAIGN.

> FOLLOW @RESERVAYLT ON SOCIAL MEDIA TO SEE UPDATES ON OUR YOUTH-LED CONSERVATION EFFORTS!

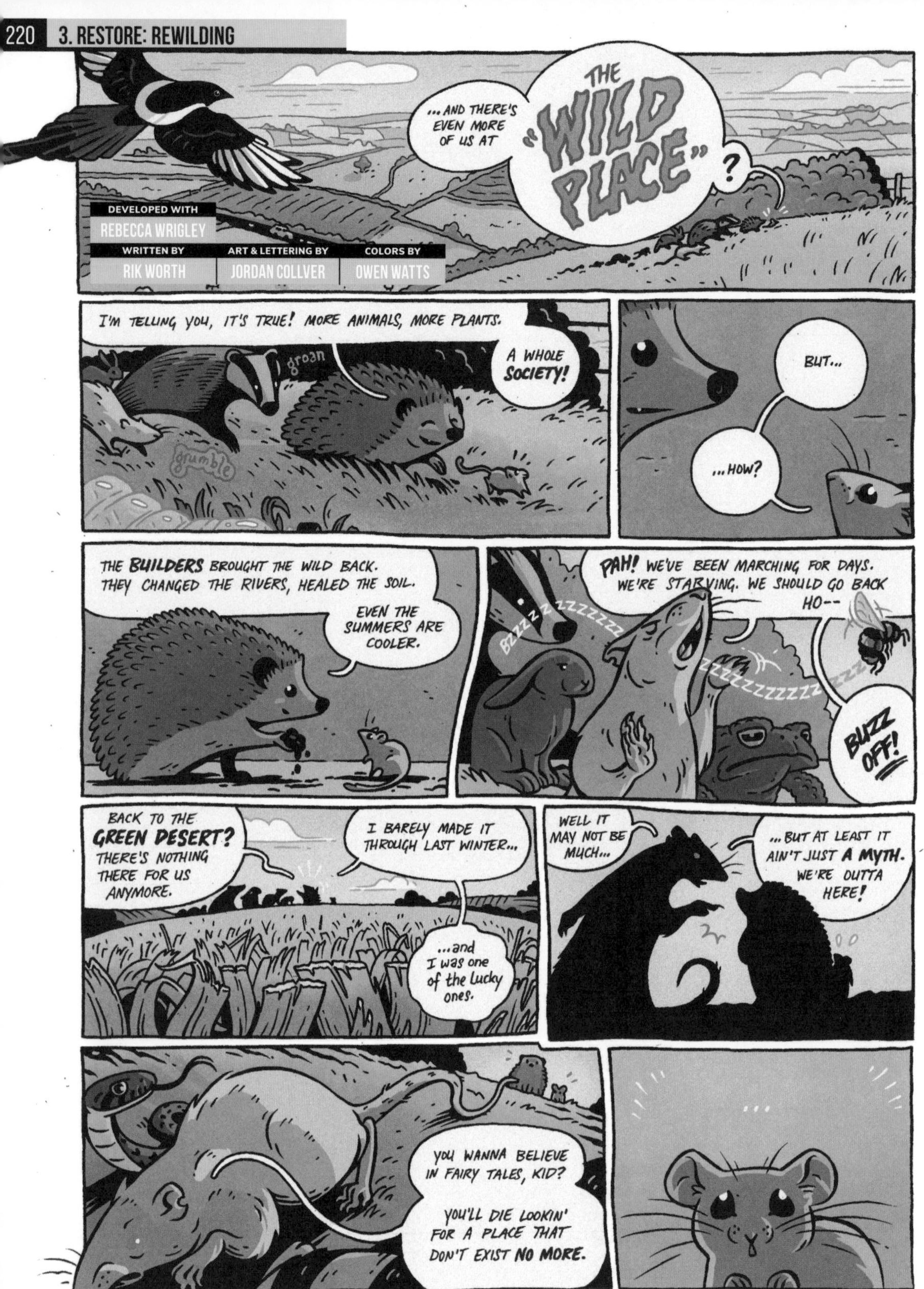
...AND THERE'S EVEN MORE OF US AT
"THE WILD PLACE"?
DEVELOPED WITH
REBECCA WRIGLEY
WRITTEN BY
RIK WORTH
ART & LETTERING BY
JORDAN COLLVER
COLORS BY
OWEN WATTS
I'M TELLING YOU, IT'S TRUE! MORE ANIMALS, MORE PLANTS.
groan
A WHOLE SOCIETY!
grumble
BUT...
...HOW?
THE BUILDERS BROUGHT THE WILD BACK. THEY CHANGED THE RIVERS, HEALED THE SOIL.
EVEN THE SUMMERS ARE COOLER.
PAH! WE'VE BEEN MARCHING FOR DAYS. WE'RE STARVING. WE SHOULD GO BACK HO--
BZZZZ ZZZZZZZ
ZZZZZZZZZZZZZZ
BUZZ OFF!
BACK TO THE GREEN DESERT? THERE'S NOTHING THERE FOR US ANYMORE.
I BARELY MADE IT THROUGH LAST WINTER...
...and I was one of the lucky ones.
WELL IT MAY NOT BE MUCH...
...BUT AT LEAST IT AIN'T JUST A MYTH. WE'RE OUTTA HERE!
YOU WANNA BELIEVE IN FAIRY TALES, KID?
YOU'LL DIE LOOKIN' FOR A PLACE THAT DON'T EXIST NO MORE.
...

I'M SORRY.
WAIT!
sniffle
I'VE BEEN SUCH A FOOL...
CHAK CHAK
OH... GOOD MORNING, MR. MAGPIE, HOW'S YOUR-
HUH?
Brree
Sniff-sniff
NEVER SEEN ANYTHING LIKE THIS BEFORE
...
IT'S GOOD!
IT'S--
GASP!
HH! HH!
HH-HH!
HH!
SNIFF
SNUF
SNIFFLE
SNUFFLE

chidi chidi
OH MY.
IT'S...
CHIRRUP-CHIRRU
IT'S REAL.
KROOOAK-KROOOAK
tiiweeee
CHAK ACKAK
THE WILD PLACE!
churr-churr chippit-chippit
seee-chik-chik
CHEWEE-CHEWEE
chii-chii
WE FOUND IT!
ABOOT TIME, EH?
tuwee Spiyu
COME... COME
MAKE YOURSELF AT HOME.
WAIT, YOU SAID "WE"... YOU EXPECTIN' MORE, EH?
I HOPE SO...
...ONE DAY SOON.

WHAT'S THE PROBLEM?

IN MANY PLACES ACROSS THE WORLD, PEOPLE ARE UNAWARE OF WHAT **THE WILD** SHOULD LOOK LIKE...

FOR EXAMPLE, MOST OF BRITAIN'S LAND HAS BEEN CULTIVATED AND SHAPED BY HUMANS TO SERVE OUR PURPOSES BUT IS **BEREFT OF LIFE!**

WILDLIFE IS CURRENTLY **DISCONNECTED**, SURVIVING IN SMALL FRAGMENTS OF LAND.

WE NEED TO EXPAND, RESTORE AND **JOIN UP** WILD PLACES SO THAT SPECIES CAN **MOVE SAFELY** -- OR WE'LL BE FACING MORE MASS EXTINCTIONS.

LET NATURE DO ITS THING!

"REWILDING" AIMS TO RESTORE ECOSYSTEMS TO THE POINT WHERE NATURE CAN TAKE CARE OF ITSELF.

WHEN IT DOES, EVERYBODY WINS!

WOODLANDS STAND STRONGER. WETLANDS THRIVE. SEAS FLOURISH. PEOPLE PROSPER.

THE WILD IS GOOD FOR PEOPLE TOO...

WE SHOULDN'T NEED TO TRAVEL TO ANOTHER COUNTRY TO BENEFIT FROM WILD PLACES!

WITH REWILDING, WE CAN CREATE NEW **NATURE-BASED ECONOMIES** AND **BEAUTIFUL EXPERIENCES** WITHIN WALKING DISTANCE OF EVERYONE'S HOMES.

FROM LEAVING PARTS OF YOUR GARDEN TO GROW WILD, TO REWILDING **30%** OF THE PLANET -- REWILDING CAN WORK **AT ALL SCALES** IN A WAY UNRIVALLED BY ANY OTHER ACTION.

WE CAN DO IT. WE KNOW HOW. WE JUST NEED TO RAISE OUR VOICES AND CALL FOR CHANGE!

PARLEY
STORY INSPIRED BY
REWILDING EUROPE
WRITTEN BY
SI SPURRIER
ART & COLORS BY
DANI
LETTERING BY
BERNARDO BRICE
I SMELL YOU.
YOU SUPPOSE YOURSELVES SECRET. DUNG-FURRED AND DOWNWIND. HUH.
GO AWAY.
GO AWAY OR THERE WILL BE TEETH.
...WE HAVE SMELLED YOU FOR HOURS. WE HAVE WATCHED YOU...FOR HOURS.
YOU ARE LIKE US, WE THINK. SOME... ECHO. SOME SHADOW. WHAT ARE YOU...?
I AM DOG. MAN-FED. BONE-TRAINED. GUARD-BRED.
AND YOU... I KNOW YOU...

YOU ARE WOLF.
STOP THAT.
SNIFF SNIFF!
NO NEED, THE FANG. NO NEED, THE CLAW. WE HAVE NOT COME FOR YOU, FRIEND. ONLY THE SOFT MEAT.
FUZZ-FURRED. HOT HEARTED AND RAW. THEY BLEED AND THEY BLEAT AND OHHHH, THE STRUGGLE! THE SNAP AND THE SPRAY. IT IS DELIGHT!
FRIEND, HUNT WITH US. THERE IS PLENTY.

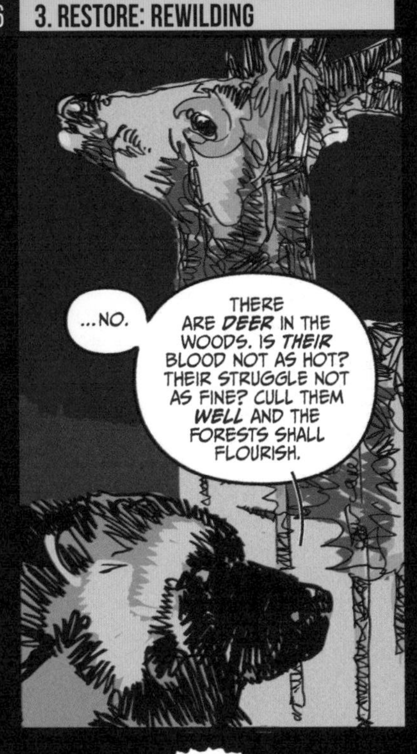
THERE ARE DEER IN THE WOODS. IS THEIR BLOOD NOT AS HOT? THEIR STRUGGLE NOT AS FINE? CULL THEM WELL AND THE FORESTS SHALL FLOURISH.
...NO.

RF.
DEER ARE HARD. FLEET LEAPERS. SWIFT OF HORN AND WARY.
WE WILL HAVE THE SOFT MEAT. THE SOFT MEAT IS EASY.

...NO.
THE SOFT MEAT IS MINE.
THE SOFT MEAT IS DEATH.

A THREAT. FOOLISH.
HM. DIVIDE, THEN. SURROUND. MY TEETH DO NOT NEED TO FIND YOU ALL.
HERE IS MY PROMISE:

I WILL TEAR OUT ONE THROAT BEFORE YOU OVERWHELM ME.

THE BIGGEST, PERHAPS. OR THE LITTLEST. IT DOESN'T MATTER.
THE PACK SHRINKS. THE SCENTS FADE. THE NIGHT LOSES A VOICE.
IS THE SOFT MEAT WORTH THAT?

WHAT DO YOU KNOW OF PACKS, LONESOME THING?!
YOU ARE MISTAKEN, DOG. THE SOFT MEAT IS THE GIFT OF THE TALL WALKERS. THEY ARE OUR FRIENDS.

"THERE HAS BEEN NO PACK HERE IN A HUNDRED MOONS, BUT WE HAVE SEEN THE SIGNS. WE HAVE FOUND THE BONES.
"THIS IS WOLF LAND. THE TALL WALKERS KNOW IT. IT WAS THEY WHO MOVED US TO THESE HILLS--DO YOU NOT SEE THAT THEY WANT US TO FLOURISH?
"DOES IT NOT FOLLOW THAT THEY WANT US TO HAVE THE SOFT MEAT?"

IT IS YOU WHO ARE MISTAKEN. THE TALL WALKERS ARE NOT ALIKE.
I KNOW THEM. THEY RAISED ME. THEY FEED ME.
HUH. PERHAPS YOU ARE SOFT MEAT, TOO...

THEN COME! TEST MY TEETH IF YOU THINK ME FEEBLE. LET US BOTH DIE! LET YOUR PACK STEAL THE SOFT MEAT AND FEAST WITHOUT YOU!

BUT KNOW THIS. NEXT TIME THEY COME?

"THUNDER. BLOOD.
"METAL JAWS IN THE EARTH AND POISON IN THEIR BELLIES."

I AM YOUR FOE. I DON'T DENY IT.
BUT WITHOUT ME TO ROAR AT YOUR SHADOWS YOU WOULD ALL BE SKINS DRYING IN THE SUN.
HM. I SUPPOSE, THEN, THAT WE ARE YOUR FOE. BUT WITHOUT US TO SING OUR SONGS AND DRIVE THE SOFT MEAT MAD WITH TERROR--
YOU WOULD HAVE NO DUTY. NO FEED.
AND NO POINT.
MOTHER... YOU SAID THE TALL WALKERS WISHED US TO LIVE WILD.
I-IS THIS WILD? IS THIS HOW WOLF IS SUPPOSED TO LIVE?
NO, LITTLE ONE. I DON'T BELIEVE IT IS.
BUT I SUPPOSE IT WORKS, IN ITS WAY.
WE WILL BE BACK.
I WILL BE WAITING. AND...
HAPPY HUNTING.

Dark Souls

Tree Hugger

"THE ABOMINABLE CHARLES CHRISTOPHER" COMIC STRIPS BY
KARL KERSCHL

£1,000,000
2,500 hectres of land rewilded.

£500,000
1,250 hectres of land rewilded.

£100,000
250 hectres of land rewilded.

£10,000
25 hectres of land rewilded.

THE MONEY WE RAISE IS HELPING:

MAKE EUROPE A WILDER PLACE

Many ecosystems, the basis of our natural wealth, are broken. But there is a way to fix it. It's called rewilding, and it's about giving nature the space to restore itself. Massive restoration of nature through rewilding will support wildlife comeback, help overcome the climate and biodiversity emergencies, and will improve our quality of life.

WHY IS THE PROJECT SO IMPORTANT?

Rewilding ecosystems across Europe is one of the best ways of tackling our current climate and biodiversity emergencies. It does not only benefit wild nature, but it also enhances the wide range of benefits that such nature gives all Europeans–from clean air and water, carbon sequestration and fertile soil, right through to flood protection, climate change resilience, and enhanced health and well-being.

We can give nature a helping hand to heal by creating the right conditions–by removing dams that are no longer needed from rivers, by allowing natural forest regeneration, and by reintroducing species that have disappeared. Then we should step back and trust nature to manage itself.

This funding will allow important wildlife species to come back, like the Iberian lynx and the Griffon vulture. These species play a critically important role and are part of the restoration of entire ecosystems.

SPECIES PROFILES

After years of decline, the European populations of some wildlife species are increasing. Such species include the lynx, vultures, and wild horses. They each play a critically important ecological role in a healthy ecosystem, yet numbers are still low. Rewilding will accelerate their recovery, boost biodiversity, and restore the important functions they play in driving many ecological processes.

> **“We need to do more than simply protect the nature we have left. We need to restore nature by rewilding large areas across the world.”**

Frans Schepers, managing director and co-founder of Rewilding Europe

CARPATHIAN MOUNTAINS, ROMANIA

One of the rewilding areas. A wilderness arc at the heart of Europe.

PROJECT IN ACTION

Rewilding is about the mass restoration of ecosystems, supporting wildlife comeback, and creating space where nature can govern itself without human intervention. Large areas of wild nature and abundant wildlife across our continent are created as inspirational showcases, benefitting both nature and people by boosting local economies where alternatives are scarce.

THE FUTURE

In the future Europe is a wilder place, with much more space for wild nature, wildlife, and natural processes shaping the landscapes. Wild nature has become a fundamental part of Europe's heritages and is an essential element in a modern, prosperous, and healthy society, creating new sources of income and pride amongst European citizens.

FIND OUT MORE

> JOIN THE REWILDING MOVEMENT! FOLLOW REWILDING EUROPE ON SOCIAL MEDIA AND SUBSCRIBE TO OUR NEWSLETTER AND STAY UPDATED ON THE LATEST REWILDING NEWS.

THE LONG VIEW
DEVELOPED WITH
SIR CHARLIE BURRELL, ISSY TREE, RUSS CARRINGTON, AND REBECCA WRIGLEY
WRITTEN BY
GEORGE MANN
ART BY
TAZIO BETTIN
COLORS BY
NICOLA RIGHI
INDUSTRIAL AGRICULTURE
AND UK FARMING
FOR POSTWAR RECOVERY

GREENING'S PESTICIDES

REWILDING BRITAIN
REWILDING AND REGENERATIVE AGRICULTURE

When it started snowing ash in Sydney it was one of the worst moments of my life...

But for all my pain and anxiety it was nothing compared to the suffering of those who were there...

After all the years of warnings... it felt like we had finally run out of time...

The Australian Wildfires undeniably proved our world is on the brink.

It felt like we were inheriting a flat-lining future.

BETTER THE DEVIL

DEVELOPED WITH
DACRE MONTGOMERY AND RE:WILD
WRITTEN BY
SARAH FLORENCE LORD
ART & COLORS BY
IAN STOPFORTH
LETTERING BY
PAUL GOODENOUGH

Australia has the worst mammal extinction in the world – we are just adding ourselves to the list.

*We need to reverse the decline. We need to rebuild the natural defences that were here **before.***

*Wildfire's start when there's a **build up** of flammable forest material.*

*But there's an animal that can help **change** that...*

*Help us **Bring the Devil Back** and stop the decline*

The Tasmanian Devil has been absent from mainland Australia for 3,000 years. But they hold the key to keeping our forests clear of brush and re-establishing our precious ecosystem.

Thank you for your love and support for our mission.
From Dacre, Sarah, Ian and Re:wild

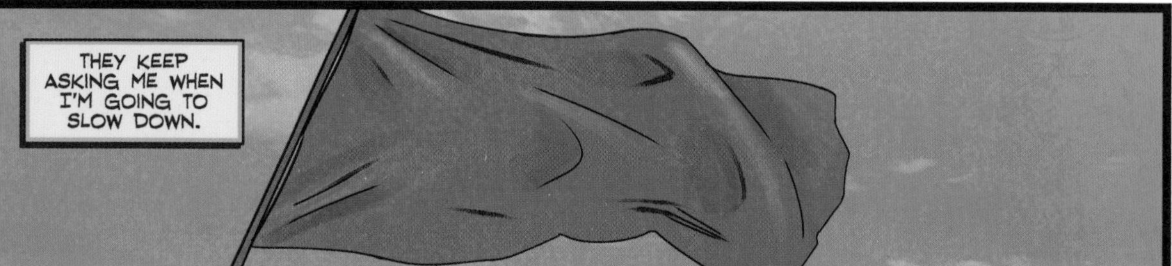

THEY KEEP ASKING ME WHEN I'M GOING TO SLOW DOWN.

NO CHAMPIONSHIP TO MY NAME. NEVER WILL BE NOW. LONG ACCEPTED THAT.
BUT THEY STILL DON'T GET IT.

Final lap for 51-year-old racer?

OH, I SEE THE SOCIAL MEDIA, TOO. *"TOO OLD FASHIONED"?* MAYBE.
75MPH
BUT THIS ISN'T THE SORT OF FAMILIARITY THAT BREEDS CONTEMPT. KINDA THE OPPOSITE.
PPM
450
350
Temp °C
1°
.7°
.4°
.1°
-.2°

AND DON'T GET ME WRONG. PLAUDITS ARE NICE. NO MATTER HOW RARE.
BUT I'M LOOKING AT SOMETHING ELSE.

WHICH LEADS TO THE FOLLOW-UP QUESTION I ALWAYS GET:
"SO... *WHY?*"

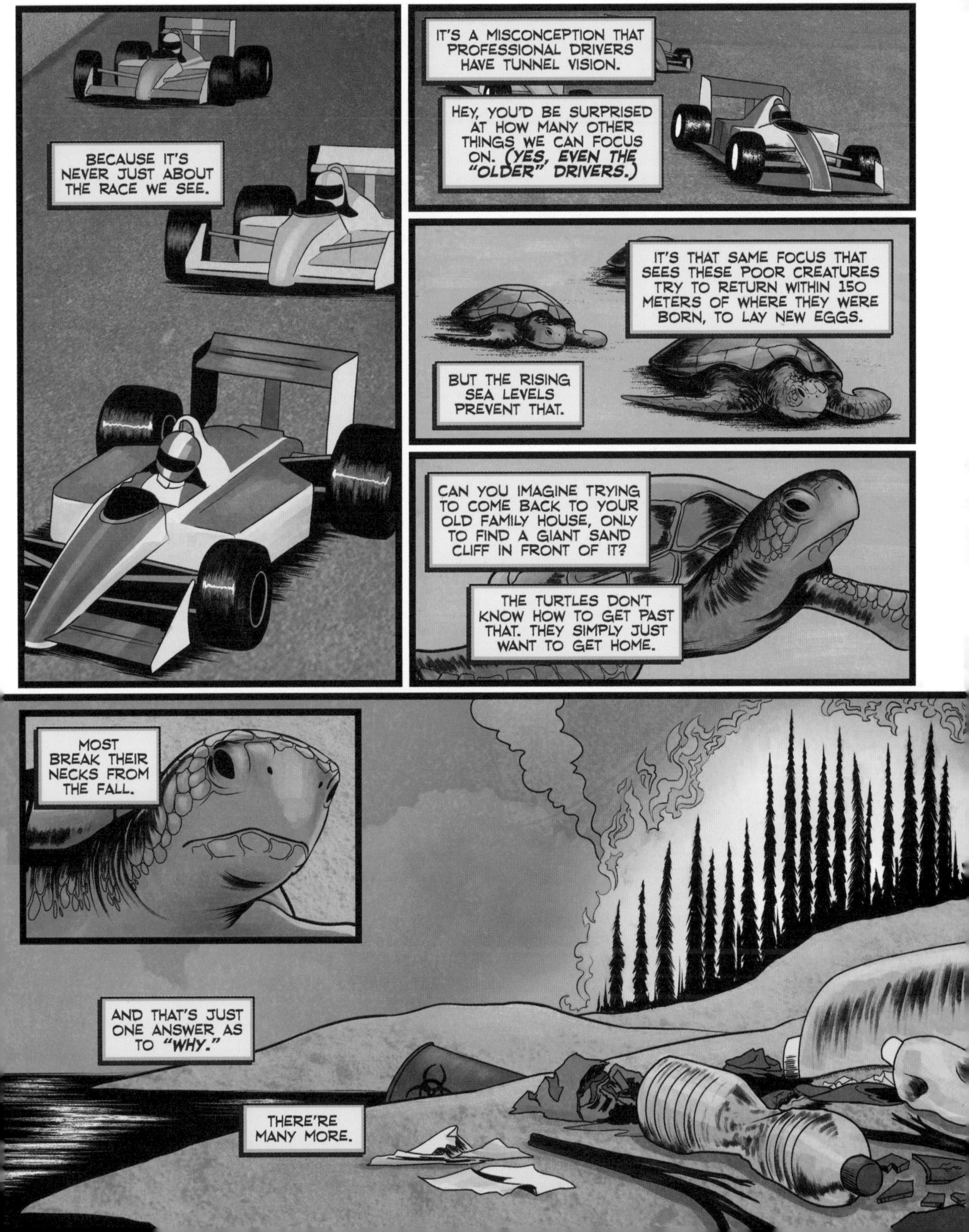

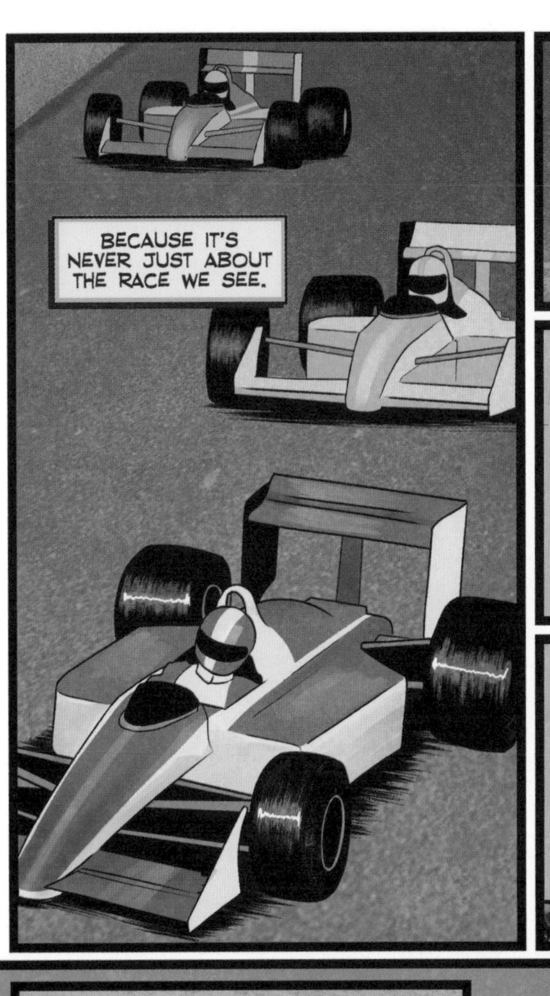

BECAUSE IT'S NEVER JUST ABOUT THE RACE WE SEE.

IT'S A MISCONCEPTION THAT PROFESSIONAL DRIVERS HAVE TUNNEL VISION.
HEY, YOU'D BE SURPRISED AT HOW MANY OTHER THINGS WE CAN FOCUS ON. *(YES, EVEN THE "OLDER" DRIVERS.)*

IT'S THAT SAME FOCUS THAT SEES THESE POOR CREATURES TRY TO RETURN WITHIN 150 METERS OF WHERE THEY WERE BORN, TO LAY NEW EGGS.
BUT THE RISING SEA LEVELS PREVENT THAT.

CAN YOU IMAGINE TRYING TO COME BACK TO YOUR OLD FAMILY HOUSE, ONLY TO FIND A GIANT SAND CLIFF IN FRONT OF IT?
THE TURTLES DON'T KNOW HOW TO GET PAST THAT. THEY SIMPLY JUST WANT TO GET HOME.

MOST BREAK THEIR NECKS FROM THE FALL.

AND THAT'S JUST ONE ANSWER AS TO ***"WHY."***

THERE'RE MANY MORE.

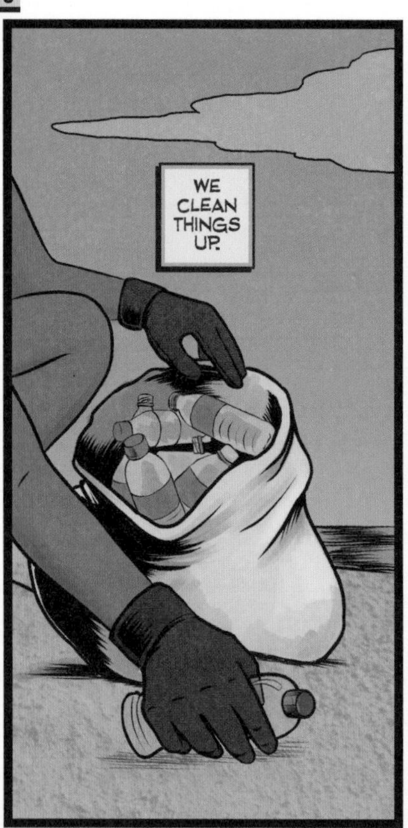

...WE RACE.

FOR SOMETHING BETTER.

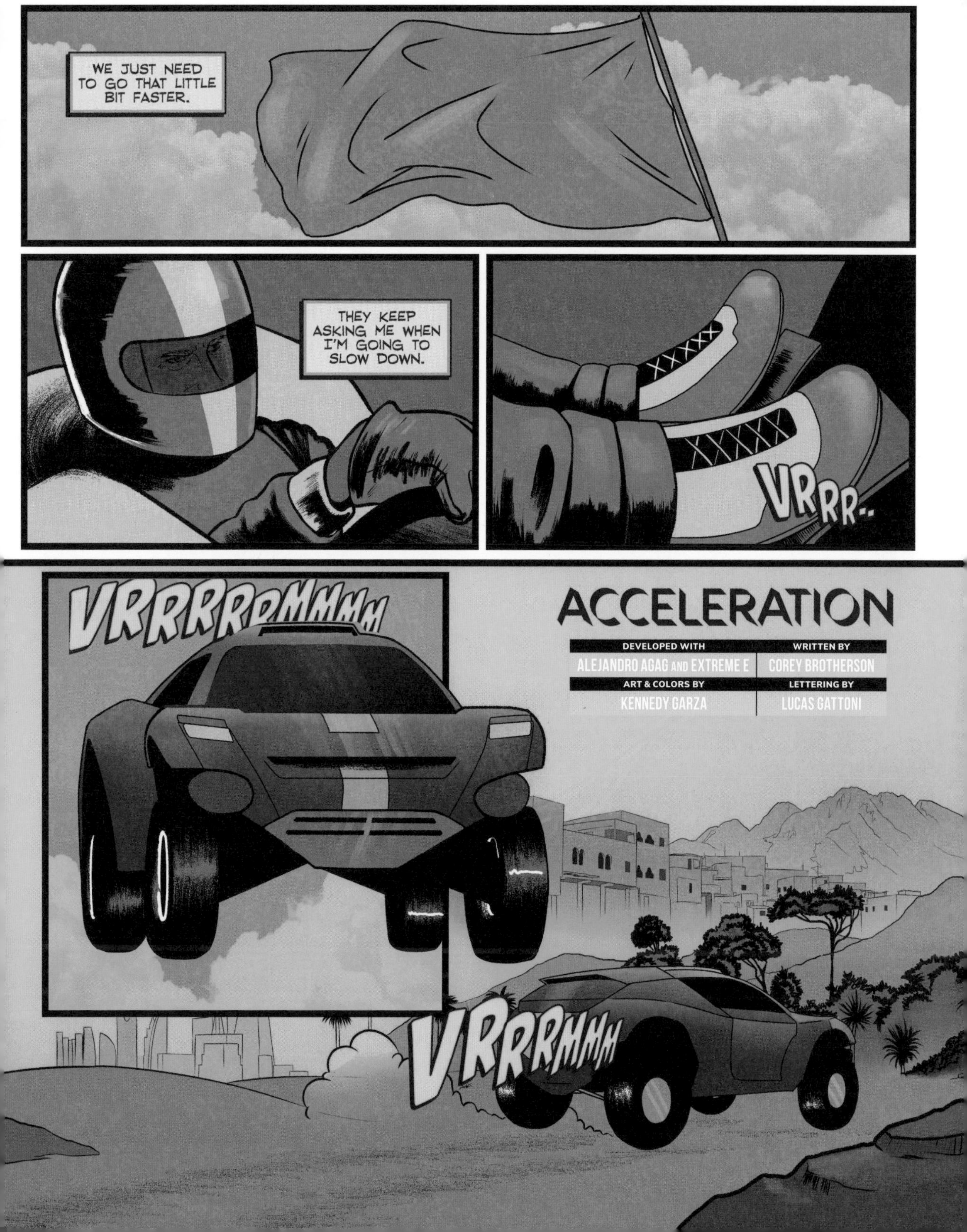

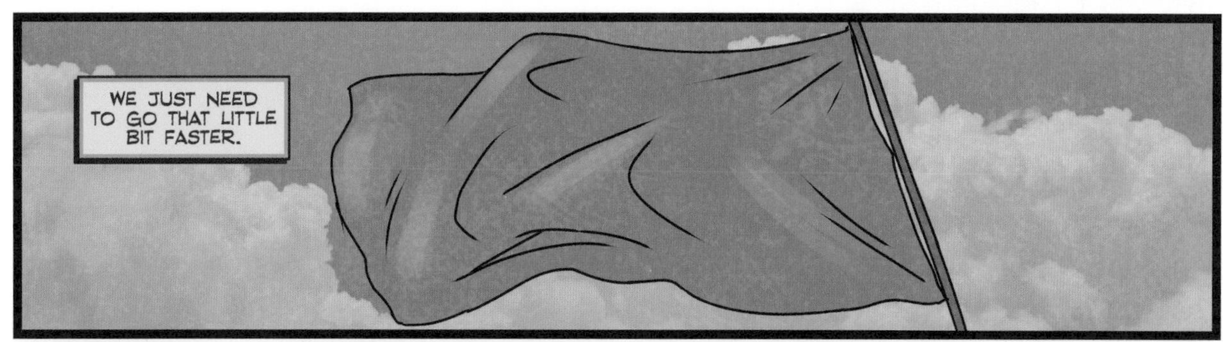

WE JUST NEED TO GO THAT LITTLE BIT FASTER.

THEY KEEP ASKING ME WHEN I'M GOING TO SLOW DOWN.

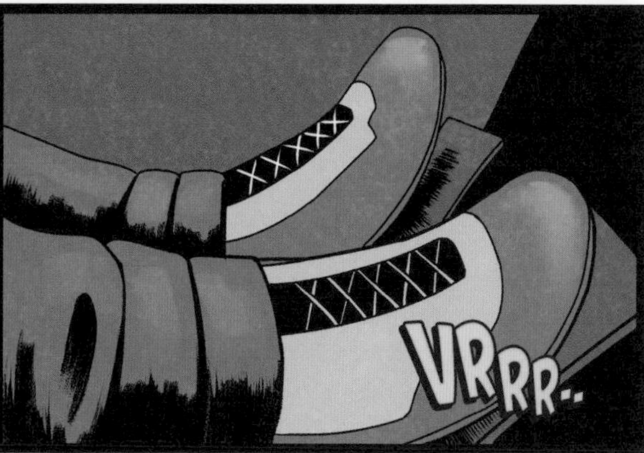

VRRRROMMMM

ACCELERATION

DEVELOPED WITH ALEJANDRO AGAG AND EXTREME E

WRITTEN BY COREY BROTHERSON

ART & COLORS BY KENNEDY GARZA

LETTERING BY LUCAS GATTONI

VRRRMMM

I'M HERE TO TELL YOU ABOUT A LIE.
NURTURE NATURE
A LIE THAT'S BEEN TOLD SINCE BEFORE YOU WERE BORN.
A LIE YOU'VE PROBABLY REPEATED YOURSELF.
THAT THESE GREAT BRITISH ROLLING HILLS ARE NATURAL, HEALTHY... WILD.

4% WILD ANIMALS
36% HUMAN
60% LIVESTOCK

BUT THE PLACES WE LOVE ARE IN TROUBLE. SOMETHING'S GONE WRONG...TERRIBLY WRONG.

WHERE WE HAVE WIND, WE ***SHOULD*** SEE THE WHITE TAIL OF THE EAGLE...

WHEN WE HEAR THE LAP OF THE WAVES, WE SHOULD ALSO HEAR THE SONG OF THE WHALE...

OUR RELATIONSHIP WITH THE PLANET IS THROTTLING WILDLIFE, CHOKING THE HUM OF LIFE...

INSECTICIDES AND PESTICIDES ARE DESTROYING THE TINY CREATURES THAT GIVE US LIFE...

IN BRITAIN ALONE, WE HAVE LOST THE BEAR AND LYNX. AND THE WILDCAT AND HEDGEHOGS ARE TEETERING ON THE EDGE...

BUT STOP, TAKE A MOMENT... AND ALLOW YOURSELF TO IMAGINE...

...A WORLD WHERE ***YOUR HOME*** IS AS WONDROUS AND ***ALIVE*** AS THE SERENGETI...

WHERE YOU CAN EXPERIENCE THE WONDER OF A SAFARI JUST BY WALKING OUT OF YOUR DOOR...

IT SOUNDS CRAZY, WE KNOW... LIKE A FANTASY. THAT'S NOT BRITAIN, YOU MAY SAY...

THAT IS THE LIE. WONDERFUL, MAJESTIC WILDLIFE WAS ***RIGHT HERE...*** AND WE ***CAN*** BRING IT BACK.

TOGETHER, WE'VE ALREADY BROUGHT BACK THE BEAVER AND THE OTTER...

...AND **RIGHT NOW** WE'RE HARD AT WORK INTRODUCING **BISON TO REWILDED WOODLANDS** AND BRINGING BACK **BUTTERFLIES** AND **FARMLAND BIRDS.**

WE **CAN** DO IT. WE **KNOW HOW,** WE JUST NEED THE WILL TO MAKE IT HAPPEN.

THE WILL TO TRANSFORM OUR NATION INTO A PLACE WHERE WILD SPACES ARE AVAILABLE TO ALL, NOT JUST THE PRIVILEGED...

...WHERE OUR TOWNS AND CITIES ARE ALIVE WITH THE MUSIC OF WILDLIFE...

WHERE THE WIND WHISTLES THROUGH THE TREES ONCE MORE AND BIRDSONG FILLS THE WOODS...

...WHERE WE GIVE 30% OF OUR LAND BACK TO THE WILD, CREATING A HYPER-CONNECTED NETWORK OF TREES AND FORESTRY TO SUSTAIN AND EMPOWER *LIFE.*

TO DO THIS, WE NEED ***YOUR HELP.*** HELP US BREAK THIS LIE AND BUILD A HOME AS ADVENTUROUS AND WILD AS ANY COUNTRY IN THE WORLD.

THE *wildlife* TRUSTS

THANK YOU.

THE GREAT BRITISH LIE

DEVELOPED WITH
THE WILDLIFE TRUSTS
WRITTEN BY
PAUL GOODENOUGH
ART & COLORS BY
SIMON MYERS
LETTERING BY
JIM CAMPBELL

WITH SPECIAL THANKS TO

DAME JUDI DENCH, BELLA LACK, JAMES MCVEY, MYA-ROSE CRAIG, LIZ BONNIN, CEL SPELLMAN, DOMINIC MONAGHAN, CHRIS PACKHAM, MEGAN MCCUBBIN, SIR IAN MCKELLEN, SIR PATRICK STEWART, MONTY DON, JEROME FLYNN, GILLIAN BURKE, PROF. BEN GARROD, TANYA BURR, JUDGE ROB RINDER, CARA DELEVINGNE, AND CRAIG BENNETT.

£5,000,000
Would help restore 10,000 hectares of precious wetland, helping myriad wildlife recover including dragonflies, otters, and marsh harriers.

£1,000,000
Would help bring back rare river woods along 100 kilometers of waterways through urban and rural landscapes.

£100,000
Would help establish a family of beavers to restore and save a wetland ecosystem.

£50,000
Would provide a toolkit to 1,000 community groups to help them campaign for nature's recovery in their area.

THE MONEY WE RAISE IS HELPING:

BRING BEAVERS BACK

The UK is one of the most nature-depleted places in the world, which is why The Wildlife Trusts have a big ambition to restore at least 30% of land and sea for nature by 2030. And beavers are key to helping us do just that! With their super engineering abilities, they inject new life into wild places, ensuring a brighter future for our wetlands and the wildlife and people that live in and around them.

WHY IS THE PROJECT SO IMPORTANT?

The next ten years will be critical in determining what kind of world we live in. We need nature for our health and well-being, food production, and to tackle the climate emergency.

It is not enough to slow the loss of wild places and the species that depend on them—we urgently need to reverse the decline and put nature into recovery. Beavers create a fantastic range of wetland habitats that provide homes for other wildlife and help address the climate emergency by locking up carbon. Not only that, the channels, dams, and wetlands that beavers engineer hold back water and release it more slowly after heavy rain, helping to reduce flooding, currently on the increase due to climate change. But these brilliant natural engineers have long been missing from Britain's landscapes, having been hunted to extinction in the 16th century. The loss of beavers led to the disappearance of the magical mosaic of lakes, meres, and boggy places that they created.

We're turning that around, and with your help, beavers will be back where they belong, creating wilder spaces that benefit us all.

SPECIES PROFILES

Kingfishers need our help because they are at risk from habitat degradation and pollution. These striking birds depend on healthy wetlands. Beavers were driven to extinction in Britain 400 years ago but are starting to return thanks to Wildlife Trust efforts. Ospreys are making a remarkable comeback. Once lost through persecution, these awe-inspiring birds fly 5,000 km back to the UK every summer to raise their chicks in our wetlands.

> "The next ten years must be a time of renewal, of rewilding our lives, of green recovery."

Craig Bennett, chief executive, The Wildlife Trusts

WETLANDS, UNITED KINGDOM

UK wetlands can be a perfect home to beavers and help store carbon, too.

PROJECT IN ACTION

Right now, The Wildlife Trusts are working with communities to secure and manage more land for nature, rewild whole landscapes, bring back lost species such as beavers, and create the green jobs of the future. We are working in partnership with local communities to reintroduce families of beavers across the UK, studying their impact on their surroundings, mapping the best possible locations, and campaigning to secure their long-term future.

THE FUTURE

We are creating a wilder future, where nature is abundant everywhere—in our towns, cities, and countryside. With more space for nature, rare wildlife will no longer cling on in isolated pockets on the brink of disappearing altogether, but thrive. Locally extinct species such as beavers have helped us bring back healthy, functioning ecosystems that sustain us and our planet. Together, with your help and support, we will get nature working again.

FIND OUT MORE

> FOLLOW THE WILDLIFE TRUSTS, AND OUR 30 BY 30 CAMPAIGN TO FIND OUT WHAT WE'RE DOING, AND HOW YOU CAN HELP.

> WILDLIFETRUSTS.ORG/30-30-30

PART OF THE REASON FOR PEOPLES' HESITATION TO MAKE CHANGES THAT BENEFIT THE ENVIRONMENT IS DUE TO THE SCALE THEY THINK IS REQUIRED.
REGENERATIONAL CHANGE
IT'S EASY TO THINK "IT'S A GLOBAL MESS--ENVIRONMENTAL ISSUES ABOUND ALL OVER THE WORLD, SO WHAT DIFFERENCE CAN ONE PERSON MAKE?"
DEVELOPED WITH GABE BROWN
WRITTEN BY CHRIS RYALL
ART BY MIKE COLLINS
COLORS BY SEBASTIAN CHENG & CANDICE HAN
LETTERING BY JIM CAMPBELL
WITH THANKS TO ARIZONA MUSE
OR "IT'S TOO MUCH--CLEAR AIR, CLEAN WATER, PESTICIDES AND FACTORY FARMING AND DESERTIFICATION OF OUR SOIL, THERE'RE SO MANY PROBLEMS...
"...AND CLIMATE CHANGE IS SUCH A COMPLICATED PROBLEM THAT NOTHING I DO MYSELF WILL MAKE ENOUGH OF A DIFFERENCE.
"IT'S NOT THAT DIRE. ENVIRONMENTAL DISASTERS LIKE THE DUST BOWL HAPPENED ALMOST 100 YEARS AGO AND WE'RE STILL GOING."
WHICH ISN'T TRUE. AND WE'RE NOT ONLY STILL REPLICATING THE PROBLEMS THAT CREATED SUCH DISASTERS AS THE DUST BOWL, WE'RE ACCELERATING THEM. MOST DROUGHTS ARE NOT NATURAL--THEY'RE MAN-MADE.
INDUSTRIAL TILLING, A LACK OF GROWING PLANTS AND BAD DECISIONS HAVE LED US TO WHERE WE ARE--TO A PRECIPICE MADE OF CRUMBLING, NUTRIENT-STARVED DIRT... FOR NOW.

BUT THE SOLUTION IS AS INTERCONNECTED AS THE WORLD ITSELF. SMALL MOVES **CAN** AND **DO** COMBINE TO MAKE BIG, POSITIVE CHANGE.
NUTRIENT-FREE DIRT WASN'T ALWAYS SO. AND IT DOESN'T NEED TO REMAIN BARREN... NOW WE HAVE THE ANSWER.
ONE THING THE PLANET DOES WELL WHEN GIVEN THE CHANCE IS REJUVENATE.
THROUGH A PROCESS CALLED **REGENERATIVE FARMING,** WE CAN TRANSFORM DIRT BACK INTO THE NUTRIENT-RICH, CARBON-ABSORBING SOIL IT ONCE WAS. WE CAN **PREVENT** ANOTHER DUST BOWL.*
*AS SEEN ON THE **KISS THE GROUND** DOCUMENTARY--ED.
REGENERATIVE AGRICULTURE HAS ALREADY BEEN USED TO RESTORE MILLIONS OF ACRES ACROSS THE WORLD...
...TURNING DESERT-LIKE LAND THAT HAD BEEN INTENSIVELY AND INDUSTRIALLY TILLED INTO INFERTILITY BACK TO THE PLANT-HEALTHY, ANIMAL-SUSTAINING SOIL IT ONCE WAS.
FARMERS WHO MAKE THIS KIND OF A CHANGE INCREASE THEIR FARM'S PROFITABILITY AND OUTPUT, AS WELL AS THE NUTRITIONAL CONTENT OF THEIR CROP YIELD.
BY PLANTING **MIXED COVER CROPS** AND GRAZING THESE WITH LIVESTOCK, FARMERS BUILD A CIRCULAR ECOSYSTEM WHERE ANIMALS FERTILIZE THE COVER CROPS WHILE GRAZING. THIS ENCOURAGES PLANTS TO DEVELOP LONGER ROOTS--WHICH **CAPTURES CARBON** RATHER THAN **RELEASING IT...**

...THIS REVITALIZATION DOES MORE THAN BENEFIT THE SOIL AND ITS CROPS: IT ALSO PROVIDES BOTH RESTORATIVE AND ECONOMIC BENEFITS TO THE SURROUNDING COMMUNITIES.
FARMERS CAN **RENEW** THE SOIL AND CREATE A MORE SUSTAINABLE ECO-SYSTEM. IT DOESN'T TAKE LONG FOR THE GROUND, AND THE LIVESTOCK THAT FEEDS OFF IT, TO SEE POSITIVE RESULTS.

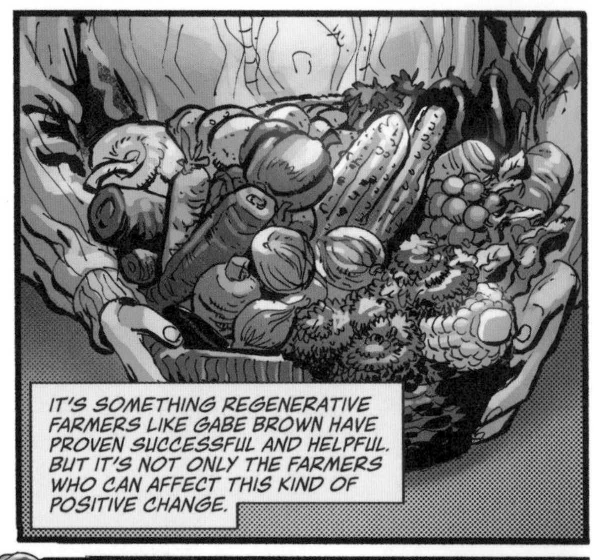
IT'S SOMETHING REGENERATIVE FARMERS LIKE GABE BROWN HAVE PROVEN SUCCESSFUL AND HELPFUL. BUT IT'S NOT ONLY THE FARMERS WHO CAN AFFECT THIS KIND OF POSITIVE CHANGE.

FIRST, THE FOODS YOU BUY AND WHERE YOU BUY THEM CAN DEMONSTRATE TO THESE FARMERS THEIR EFFORTS ARE NOT JUST WORTHWHILE BUT **SUSTAINABLE** FOR THEM.
ALSO, ANY TREES YOU CAN PLANT COMPLEMENT THE MANY OTHER EFFORTS TO REDUCE CARBON IN THE ATMOSPHERE. IT DOESN'T TAKE MUCH, AND THE ENVIRONMENTAL RELIEF IT BRINGS IS EXPONENTIAL...**WORLD CHANGING.**

AND YOUR OWN GARDENS--LARGE OR SMALL, BACKYARD OR WINDOWSILL--CAN FOLLOW THESE SAME SIMPLE STEPS TO SOIL HEALTH.

6 PRINCIPLES

Of Soil Health

THE VEGETATION DIVERSITY LEADS TO GREATER BIODIVERSITY AND PROVIDES NUTRIENT-RICH SOIL WITH REAL MEANS TO WARD OFF PREDATORY PESTS.

REDUCED PESTS MEAN LESS DAMAGE TO THE CROPS AND REDUCED NEED FOR SOIL- AND CROP-DAMAGING PESTICIDES.

1

Know your context.
Our soil health practices are a reflection of ourselves and our stewardship of the land.

2

Do not disturb.
In nature, there is no mechanical or chemical disturbance.

3

Cover and build surface armor
to protect the soil's "skin."

4

Mix it up
with a diversity of plants, microbes, insects, wildlife, livestock. Mother Nature did not grow monocultures so why should we?

5

Keep living roots in the soil
as long as possible each year. Roots feed soil microorganisms, which feed our plants.

6

Grow healthy animals and soil together.
Grazing has been an essential component of all soils at one time or another.

brought to you by SOIL HEALTH ACADEMY

REGENERATIVE FARMING IS AS IMPORTANT AS IT IS EFFECTIVE. AND IT'S SOMETHING EVERYONE CAN HELP WITH AND PARTICIPATE IN, FARMER OR URBAN-DWELLER.

PLANTS AND TREES ARE EFFECTIVE AT PULLING CARBON FROM THE ATMOSPHERE AND PLACING IT INTO THE SOIL, WHICH IS EQUIPPED TO PROCESS IT AND HOLD IT.

FROM THERE, A HEALTHY, CARBON-RICH BED OF SOIL HELPS INCREASE THE QUALITY OF LIFE OF THE PLANET AND THE PEOPLE ON IT.

IT'S NOT COMPLICATED TO DO. ALL IT REQUIRES IS GETTING YOUR HANDS A BIT DIRTY AND LETTING THE SOIL DO THE REST.

i do what
i can
i can whenever
i can
whenever

A Fevered Mind and a Pot of Pee

a sad but true account from Lucy Lawless

I learned that one of the important things when planting is learning how to produce **biochar**.

It's basically charcoal from wood burnt at very high heat with little oxygen, then doused with water and "activated" by being soaked in nitrogen for three weeks.

You **can** use slurry for the nitrogen, but I didn't exactly have any.

There is one thing that can do the job nicely... **pee!** And I can make that alllll day... and I even have **some** control over the volume...

And I found this old beaten-up pot amongst the bushes. "That'll do," I thought.

So yeah, I've been living on the farm, **sorry, I should say "my home" now...** With no landline, no cell service and no wi-fi for **months**.

It doesn't feel very safe... so not really the best time to get a head cold, you might say...

*So what do you do?
You remember why you do
these things, right?*

*"Why you'd risk your career
and leave Bel Air to buy a
farm to rewild?", they ask.*

*Well... because I can!
I'm lucky enough to have the
privilege of choice... and
once you know you can help,
how can you not?*

*So I do what I can,
whenever I can...
because too many people
have been screwed over
for far too long.*

*Here in New Zealand there's
this deal, converting sheep
farmers to dairy farmers...
and it's* ***impoverishing*** *our
farmers, our soil, our water
and our animals.*

***They made more money in
the '90s!*** *Before they were
suckered into this intensive
farming BS.*

Now, they're working
harder *than ever to
make* ***less*** *money, and
they're inadvertently
destroying our country
and our environment too!*

It's got to stop.

banj oak trees
you plant them
and you've got
a whole economy

The other day, a young man named Nigel knocked on my door.

He said to me, "I see your trees are coming on"...

In my headcold-induced psychosis, I said something to the effect of:

"Can you believe what's going on?! They're planting these $*%#&@ pine forests everywhere and they make really $*%#&@ timber and after two or three times planting those trees the soil's bloody **useless!** **Banj Oak** trees... that's the answer! You plant them instead and you've got a whole **economy**! The animals love their acorns, the timber from it is super thick and it stores a $*%#&@ ton of carbon AND you can grow all sorts of stuff on their bases!"

Poor Nigel's response?!

"I'm umm... just here to set-up your wi-fi, Miss Lawless..."

This damned sickness seems never ending.
But I just keeping going. Planting. Peeing. Planting. Peeing...
And just when I think it will never end...
It does...just in time. I just needed to keep going.
Now if you'll excuse me, I have some more 'business' to attend to.
$%&@...where did I leave that pot?!
WRITTEN BY
LUCY LAWLESS, PAUL GOODENOUGH & TYLER JENNES
ART BY
DAVID MACK
LETTERING BY
PAUL GOODENOUGH
MACK

GROW YOUR OWN, SAVE THE WORLD

PRODUCING FOOD CLOSER TO WHERE IT'S EATEN MEANS LESS IS SHIPPED AROUND THE WORLD. BECAUSE BURNING FUEL TO TRANSPORT OUR FOOD IS JUST MAKING THE CLIMATE CRISIS WORSE.
IT'S ALSO FRESHER, TASTIER AND MORE NUTRITIOUS. DAMN SATISFYING, TOO.
AROUND THE WORLD, WE GROW FAR MORE FOOD THAN WE NEED. ABOUT ONE THIRD IS LOST BETWEEN THE FARM AND OUR PLATES. ANOTHER THIRD IS FED TO ANIMALS TO PRODUCE MEAT AND DAIRY FOODS.
SO MOST OF THE CROPS WE GROW AREN'T EATEN BY PEOPLE. WHEN MILLIONS ARE STILL GOING HUNGRY, WASTING SO MUCH FOOD IS CRIMINAL.
DIGGING, PLANTING AND HARVESTING ARE GOOD FOR YOUR MIND AND BODY.
IT KEEPS YOU ACTIVE, AND STUDIES SHOW IT HELPS YOUR MENTAL HEALTH AS WELL.
INTENSIVE FARMING IS WIPING OUT SPECIES, DAMAGING SOILS AND FUELING CLIMATE CHANGE, SO WE HAVE TO CHANGE HOW OUR FOOD IS PRODUCED.
HERE ARE SOME OF THE THINGS YOU CAN DO TO HELP.
- Grow your own at home or in a community garden.
- Buy local, seasonal and organic food if you can.
- Ask your local government to make space for community farming.

DEVELOPED WITH PROF. DAVID GOULSON | WRITTEN BY JAMIE WOOLLEY | ART BY GORAN GLIGOVIC | LETTERING BY BERNARDO BRICE

THE STORM MUST HAVE BLOWN A HOLE IN THE FENCE. JUST LOOK...
Untidy
DEVELOPED WITH
DOMINIC MONAGHAN
WRITTEN BY
PAUL GOODENOUGH
ART & COLORS BY
GEOFF SENIOR
LETTERING BY
JIM CAMPBELL

DAAAVIIIID!
YOU BEST NOT BE MESSING AROUND ON MY LAWN AGAIN. I'M BLOOMING WARNIN' YOU!

HOW MANY TIMES DO I HAVE TO TELL YOU?
MY GARDEN ISN'T FOR MESSING ABOUT IN! NOW GET IN HERE BEFO--

...DAD?

CALL AN AMBULANCE, DAVID!
QUICK!

AS THE AMBULANCE TOOK MY DAD AWAY, I COULD STILL HEAR HIM MUMBLING...
"GET THOSE... BLOODY DIRTY SHOES OFF MY FLOOR..."

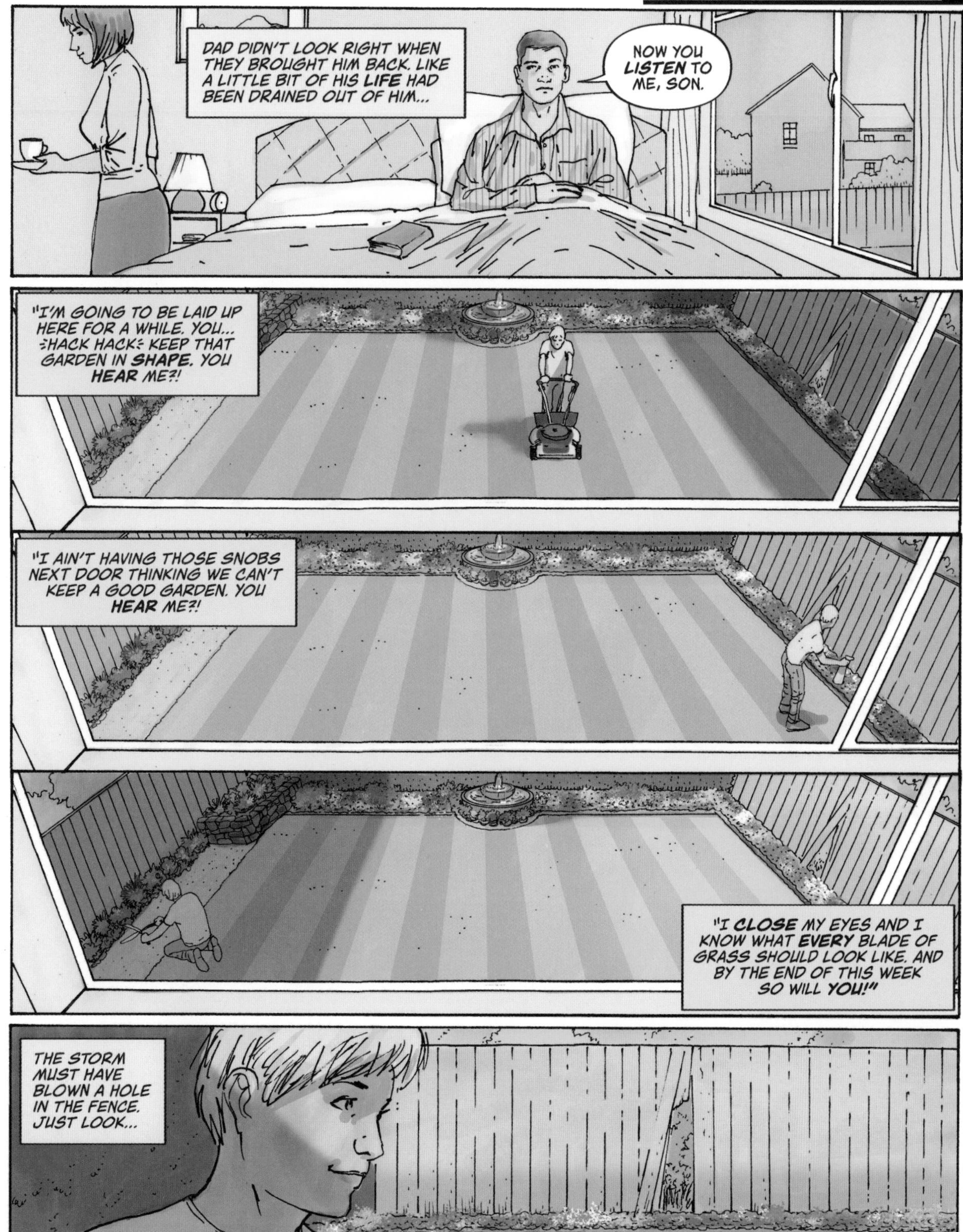
DAD DIDN'T LOOK RIGHT WHEN THEY BROUGHT HIM BACK. LIKE A LITTLE BIT OF HIS LIFE HAD BEEN DRAINED OUT OF HIM...
NOW YOU LISTEN TO ME, SON.
"I'M GOING TO BE LAID UP HERE FOR A WHILE. YOU... ⫶HACK HACK⫶ KEEP THAT GARDEN IN SHAPE. YOU HEAR ME?!
"I AIN'T HAVING THOSE SNOBS NEXT DOOR THINKING WE CAN'T KEEP A GOOD GARDEN. YOU HEAR ME?!
"I CLOSE MY EYES AND I KNOW WHAT EVERY BLADE OF GRASS SHOULD LOOK LIKE. AND BY THE END OF THIS WEEK SO WILL YOU!"
THE STORM MUST HAVE BLOWN A HOLE IN THE FENCE. JUST LOOK...

I KNEW IF I WAITED FOR EVEN A SINGLE HEARTBEAT MY COURAGE'D BE GONE!

SO I SHUT MY EYES, HOPED MY DAD WAS ASLEEP AND--

--I STEPPED THROUGH.

I'D NEVER SEEN THE OIL FACTORY IN REAL LIFE. ONLY PICTURES. AND THEY DIDN'T LOOK **ANYTHING** LIKE THIS!

AND INSIDE THERE WAS A WHOLE OTHER WORLD.
ANTS, WOODLICE, MILLIPEDES...IS THAT FUNGAL HYPHAE?
NEMATODE WORMS, SPRINGTAILS, SILVERFISH...

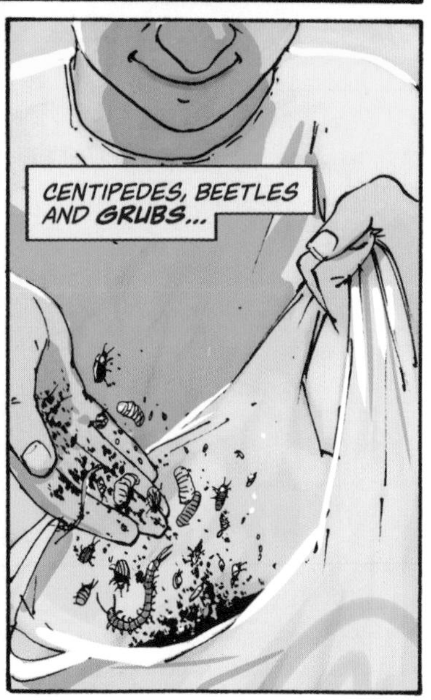

CENTIPEDES, BEETLES AND GRUBS...

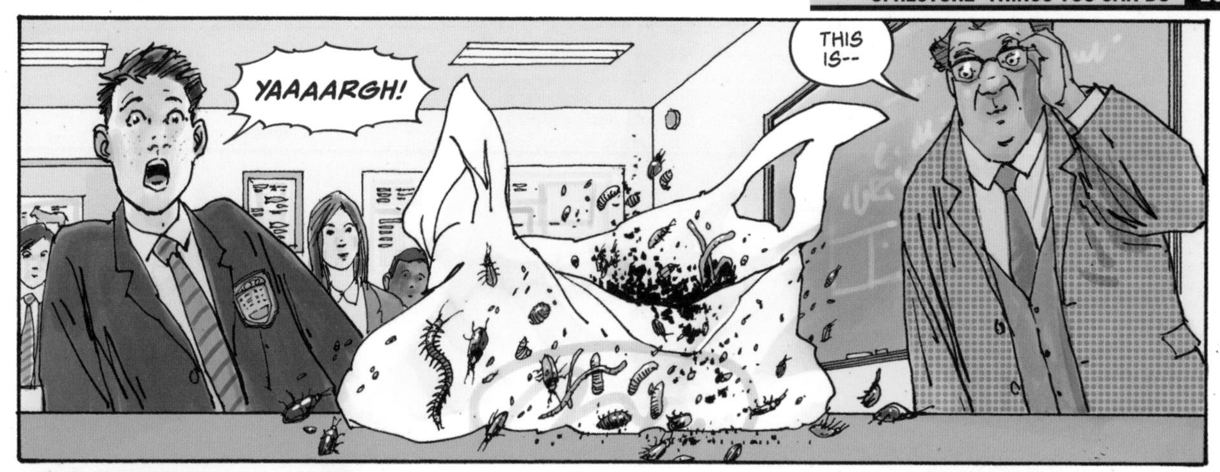
YAAAARGH!
THIS IS--

--JUST...UNBELIEVABLE! SOME OF THESE ARE REALLY RARE! WHERE ON EARTH DID YOU FIND THEM?
THE OLD...OIL FACTORY?
NO ONE'S BEEN NEAR THAT PLACE FOR YEARS. ALL OF THESE COME FROM THERE?!? THIS IS INCREDIBLE, DAVID. YOU SHOULD TELL YOUR PARENTS! YOU'VE MADE A GENUINELY GREAT FIND.

WHATTT?!!

ARE YOU JOKING, LAD? WHY DO YOU THINK WE SPRAY MY GARDEN WITH PESTICIDES? TO GET RID OF BLEEDIN' BUGS LIKE THESE!?

GO ON-- CLEAR OFF! AND TAKE THEM FILTHY BUGS WITH YOU, AN' ALL!

JEN!! WHERE'S THAT =HACK HACK HACK= WHERE'S THAT SON OF OURS?

SOLDIER BEETLE, LACEWING, PILL BUG...MOUSE SPIDER... DUNNO WHAT THAT ONE'S CALLED...

LOOK AT IT OUT THERE! I KNOW HE'S BRINGING THAT...MESS IN FROM THE FACTORY. I DUNNO WHY. IT'S HORRIBLE! BLOODY HORRIBLE...
IT'S JUST... DIFFERENT!

EXACTLY! DIFF'ERENT! I LIKE IT ALL THE SAME. CLEAN. PERFECT! THIS...MESS IS--
YOU HAVE A WORD WITH THAT BOY OR I =HACK HACK HACK= I WILL.

WHERE'S HE GOING THIS TIME, THAT LITTLE--
=COUGH= =COUGH=

WHEN I GET MY HANDS ON THAT KID I'M GOING TO-- =COUGH= =COUGH=
PEST-BE-DEAD

...
PEST-BE-DEAD

...WHO ARE YOU LOT?
WOULD YOU LOOK AT THAT. AIN'T SEEN A ROBIN SINCE I WAS A KID...
HONEY? WHERE ARE--
SHHH! I JUST SAW SOMETHING.
GET THEM BINOCULARS. HURRY!
THERE HE IS. LITTLE BLIGHTER!
IS EVERYTHING... OKAY?
PAUL?! DID YOU SEE SOMETHING?
YEAH. YEAH I DID.

OI! GET IN HERE!

NOW...WHAT YOU GOT THERE?

I'M SORRY, DAD. I WAS GOING TO... I WAS GOING TO... PUT IT INTO OUR GARDEN.

OH RIGHT! OUR GARDEN, IS IT?!?
OURS...?

HONEY...?
DAVID?!
"I WAS SO ANGRY WHEN I FIGURED OUT WHAT YOU WERE DOING."

YOU SEE...I WAS BROUGHT UP TO LIKE THINGS TIDY...AN' I HAD THIS PICTURE IN MY HEAD-- I KNEW EXACTLY WHAT I WANTED MY GARDEN TO LOOK LIKE.
AND THEN *YOU* WENT AND MESSED IT ALL UP!

AND SOMEHOW, THIS GARDEN OF... ***OURS,*** WENT FROM BEING THIS PERFECT ***PICTURE...*** INTO LIKE... A STORY OR A FILM OR SOMETHING...
...AND EVERY DAY WAS DIFFERENT, ***EVERY DAY...***

...GAVE ME A REASON TO GET UP. WELL, ASIDE FROM SCREAMING AT YOU. I'VE NEVER BREATHED SO ***WELL*** IN ME LIFE!

COME ON THEN...LET'S HAVE A LOOK AT OUR NEW ***GUESTS!***

"Rewilding is not about abandoning civilization but about enhancing it."
—George Monbiot

CHAPTER FOUR

INSPIRE

AND EDUCATE

The planet is in crisis and the ability for positive change lies with us all. How we appreciate and interact with the natural world, and the myriad species that call it their home, is now mission-critical.

Every single one of us can play a part in creating a better future and, to inspire us each step of the way, we've brought together some of the most influential voices on the planet to share their thoughts, fears, hopes, and dreams — and tell stories to help us all think, engage, be empowered, and take action.

I Love You, Earth

I love you, earth, you are beautiful I love the way you are
I know I never said it to you
But I wanna say it now

I love you, I Love you I love you, earth
I love you, I love you I love you nOw

I loVe you, earth, you are beautiful
I love the way you shine
I love your valleys, I love your mornings In fact I love you Everyday

I know I never said it to You
Why I'd never know
Over blue mountains, over green fields I wanna scream about it nOw
I love you, I love you I love yoU, earth
I love you, I love you I lovE you now
You Are our meeting point of infinity You are our turning point in eteRnity

I love you, I love you (I love you, I love you) I love you, earTh
I love you, I love you (I love you, I love you) I love you now
I love you, I love you (I love you, I love you) I love you, eartH
I love you, I love you (I love you, I love you) I love you now

Timo Wuerz 2021

YOKO ONO

IN PARTNERSHIP WITH
LETTERS TO THE EARTH

INSPIRED AND WRITTEN BY
YOKO ONO
ART & LETTERING BY
TIMO WUERTZ

USE YOUR VOICE

Dear Parliamentarians,

I have been a nurse for forty years;
I have worked for the last ten with people with dementia.
Dementia is no respecter of status, wealth, class, or prestige.
I have nursed doctors, lorry drivers, and paedophiles.
I accord them all the same human dignity and respect;
I give them appropriate care regardless of who they are and how they have lived their lives.

I have learned to communicate with those whom dementia has
robbed of their ability to speak coherently;
I have learned to communicate with those who are mute,
literally having no voice.
You have a voice, and you have the status and power
to use your voices to represent those of us with none.

My five-month-old grandson does not yet have words; he communicates his joy at life in myriad ways.
He laughs, he squeals, his smile beams like a beacon on a stormy day.
His joy is palpable; he wants to experience life in all its beauty and richness.
But if climate change continues as predicted by the IPCC
what will be his future?
And what of the future for your children and grandchildren?

Please, on behalf of all of those with no voice,
use yours to bring change, so we may all have a future.

Yours faithfully,
Marian Greaves RGN/Mother/Grandmother

Illustration by Dan Bradbury
In partnership with Letters To The Earth

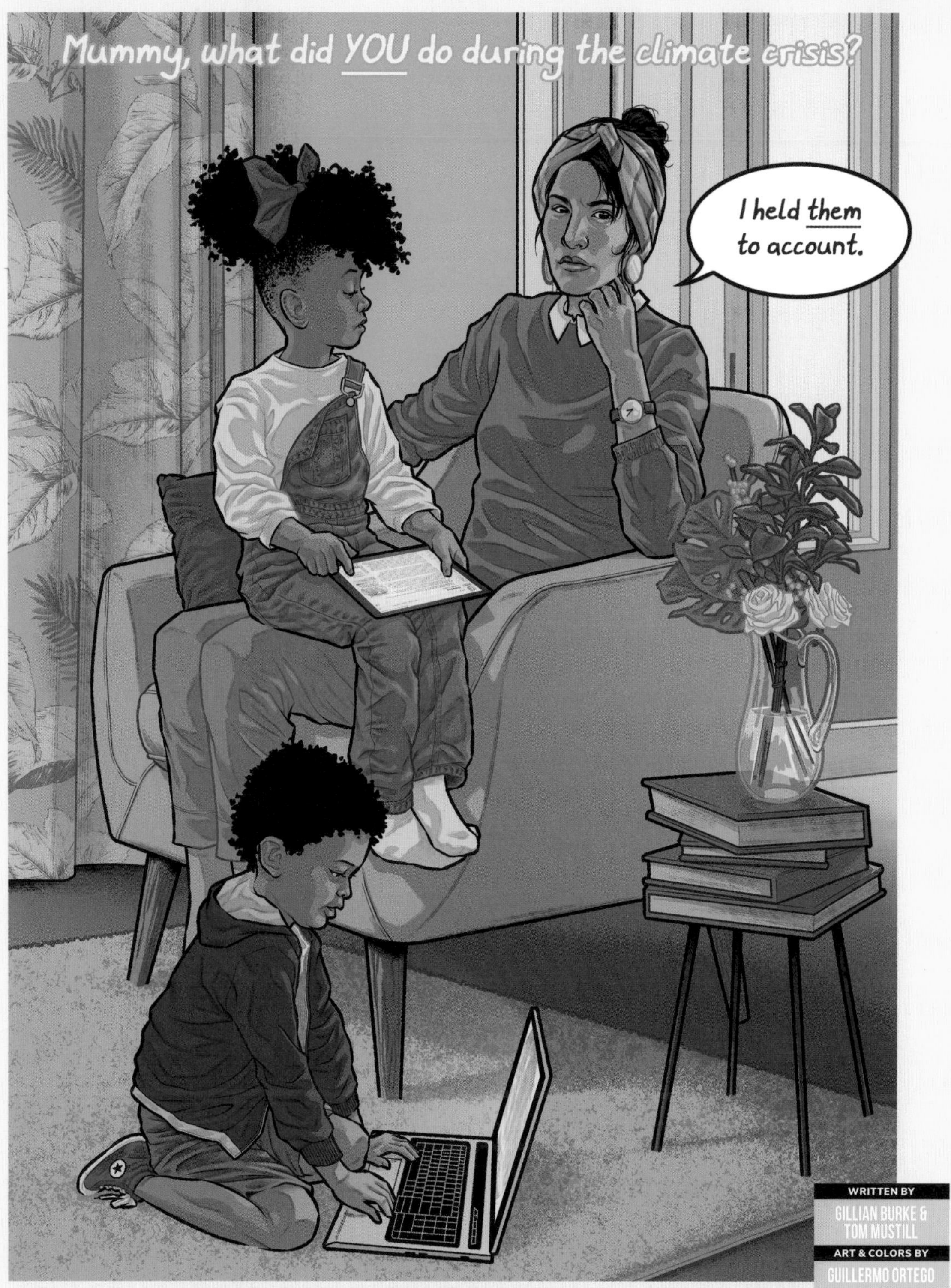

Just 3 companies emit 3 times more carbon than all UK households combined. You're doing your bit. It's time for them to do theirs.

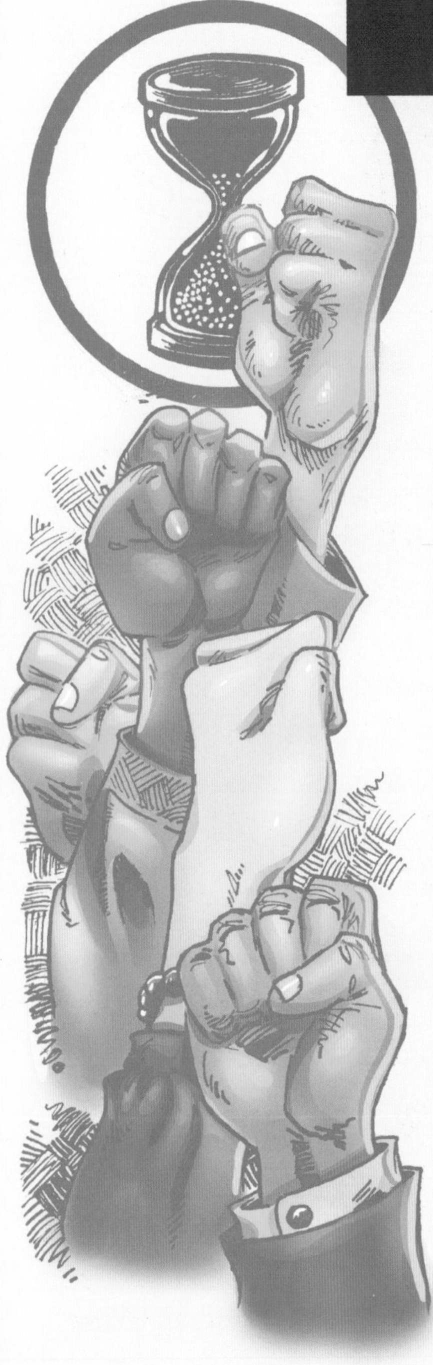

INTERVIEW WITH EXTINCTION REBELLION CO-FOUNDER

DR. GAIL BRADBROOK

Interview by John Wagner, Art by Abdulkareem Baba Aminu

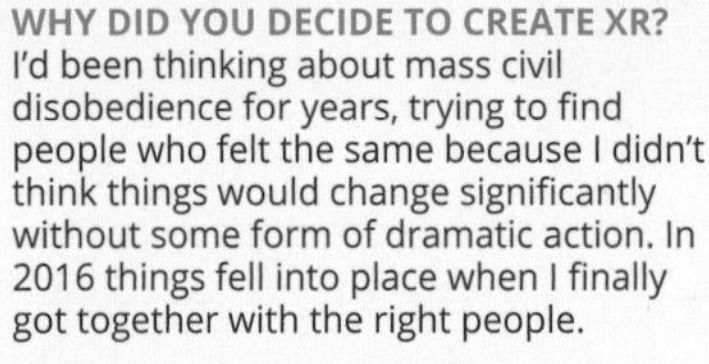

WHY DID YOU DECIDE TO CREATE XR?
I'd been thinking about mass civil disobedience for years, trying to find people who felt the same because I didn't think things would change significantly without some form of dramatic action. In 2016 things fell into place when I finally got together with the right people.

HOW DID YOU SET ABOUT IT?
We did lots of research and took training in how to build social movements. We developed a talk and training aid and every time someone "liked" our Facebook page we contacted them to ask if they would organize a local talk. We already had a base, building from 2014–2018 through work I did with Compassionate Revolution and then together with a group we called Rising Up. We tried lots of small actions. It took us 25 iterations to think of the name Extinction Rebellion and some of us nearly fell out over it!

WHAT IS YOUR ATTITUDE TO THE DISRUPTION XR CAUSES?
On the one hand building a garden bridge in Waterloo and parking a boat in Oxford Circus, setting up a community at Marble Arch, they were glory days of togetherness and possibility, of saying "screw you" to a system that is killing life on Earth.

We fully understand that we can mess up people's days, especially if they are trying to get to see a loved one or deal with another important matter, and that's upsetting. It doesn't feel good inconveniencing people, stopping them going about their daily business, having them caught up in our actions. We know they can be indiscriminate in their targeting.

We do make some people angry, but my experience is many are resigned and accept the necessity of what we're doing. Often they thank us.

WHAT ARE YOUR MAIN AIMS RIGHT NOW?
For myself and the Money Rebellion team I am working with—to point out that we have failed to tackle this crisis because we have a political economy with destruction baked in.

We want to create a global citizens' assembly to help rewire humanity and challenge the fairly hidden power holding bodies, like those that set international tax rules, or are wedded to economic growth. Above all we want to keep up the pressure on our governments. COP 26 is a good place to get our point across.

HOW DO YOU FEEL THE MOVEMENT SPREAD?
Because we had clear demands, tactics, a theory of change and principles and values—it was relatively easy to get involved and there was a zeitgeist, a wave that this built from, from the work of others: the IPCC report, Deep Adaptation (a social movement developed by Jem Bendell), Greta Thunberg, David Attenborough, the Sunrise Movement, etc.

DO YOU BELIEVE YOU CAN SUCCEED?
We can't—we have already failed. Climate tipping points are being breached now, there's no carbon budget left. We are managing a decline and a disaster that will kill millions, probably billions. We are living through the time when the biggest crimes against humanity are being committed.

Whatever success we have will be in living well, with integrity, and in working together to bring as much mitigation as we can to the climate disaster, in saving as much of our world as we can. It feels like we're hanging on to the possibility of a miracle—a great awakening and the best of humanity at last focusing on healing the damage we have done. It will require a global effort on a scale that's hard to imagine.

IF YOU COULD CHANGE ONE THING, WHAT WOULD IT BE?
That those with the most access to power in the world would wake up, collaborate and let go of the attachments they have to power and money, and go on a massive adventure of reimagining what is possible. I would like to see people like Rupert Murdoch changing their ways, to see billionaires using their money to fund the transition, to see the rapid demise of the fossil fuel industry.

And for people, all people to feel inspired to live well, with purpose and actions… Life wants us to live… It's a joyful thing to do, but you have to be willing to be transformed away from grief and to see your own power as the bright light that it is, to face fear and never give up, to find others like you and join forces.

ClimateTBD

WRITTEN AND DRAWN BY
ROSEMARY MOSCO

THE PEOPLE GATHERED FOR THE TIME-HONORED CEREMONY
TO WELCOME THIS PRECIOUS NEWBORN INTO THEIR COMMUNITY.

AND AFTER ALL THE WORDS OF GRATITUDE FOR THE SEEN AND UNSEEN BEINGS THAT MADE LIFE SO WORTH LIVING HAD BEEN SPOKEN, A HUSH FELL UPON THEM.

SLOWLY, THE ELDER RAISED THEIR CLEAR EYES FROM THE FIRE, SWEPT BACK THEIR LONG HAIR, AND FROM BENEATH SMILE LINES ETCHED IN SHADOW, SPOKE:

It is time for each of us to be reminded of the story that once was forgotten, and shall not be again, for it is a truth that has held our village through these past centuries and allowed us to thrive.

For it is said that back then, a great dread had taken hold... It seemed, they say, that all would surely be lost—and yet, because of this simple fire, there was a remembering of what long-distant cultures the world over had always known before the dread began.

This is how it was told to me:

The grandmothers noticed there was a need for the village council to be held in the light of a force greater than their own individual desires.

It is said that they sought guidance from the spirits of the directions, of the plants, and of the animals. They consulted with their ancestors, with the unborn generations and with the spirits of the land.

It is said that the spirit of fire offered itself in service. And so it was that a grandmother kindled a little fire and placed that flame at the centre of the council and had each one present make this pledge to that fire.

No decision, no thought, no action of any kind will emanate from this council unless it holds the good of the children of all beings, of this generation and seven generations hence, at its core.

And it is said that the making of this one simple pledge, held clean and pure by the transformative power of fire, marked the turning point, bringing life back to the people—and to all beings.

It is incumbent upon each of us to remember, in all that we do, to embody this pledge—to tend this fire by living our lives in honor of the children, and in this way, we too will become good ancestors.

AND SO, WITH THESE WORDS, I WELCOME YOU, PRECIOUS CHILD INTO THIS WORLD.

The Children's Fire, from those who would call us their ancestors.

Written by David Smart Knight and Jerome Flynn. Art by William Simpson.
In honor of the wisdom of our Indigenous Brothers and Sisters.

www.childrensfire.earth

REWILD FOR VICTORY

CONCEPT BY
TOM MUSTILL

POSTER DESIGNED BY
MAGGIE BEHLING

ASK NOT ONLY WHAT YOU COULD DO FOR YOUR COUNTRYSIDE,

ASK WHAT YOUR COUNTRYSIDE COULD DO FOR YOU.

Future Archaeologist

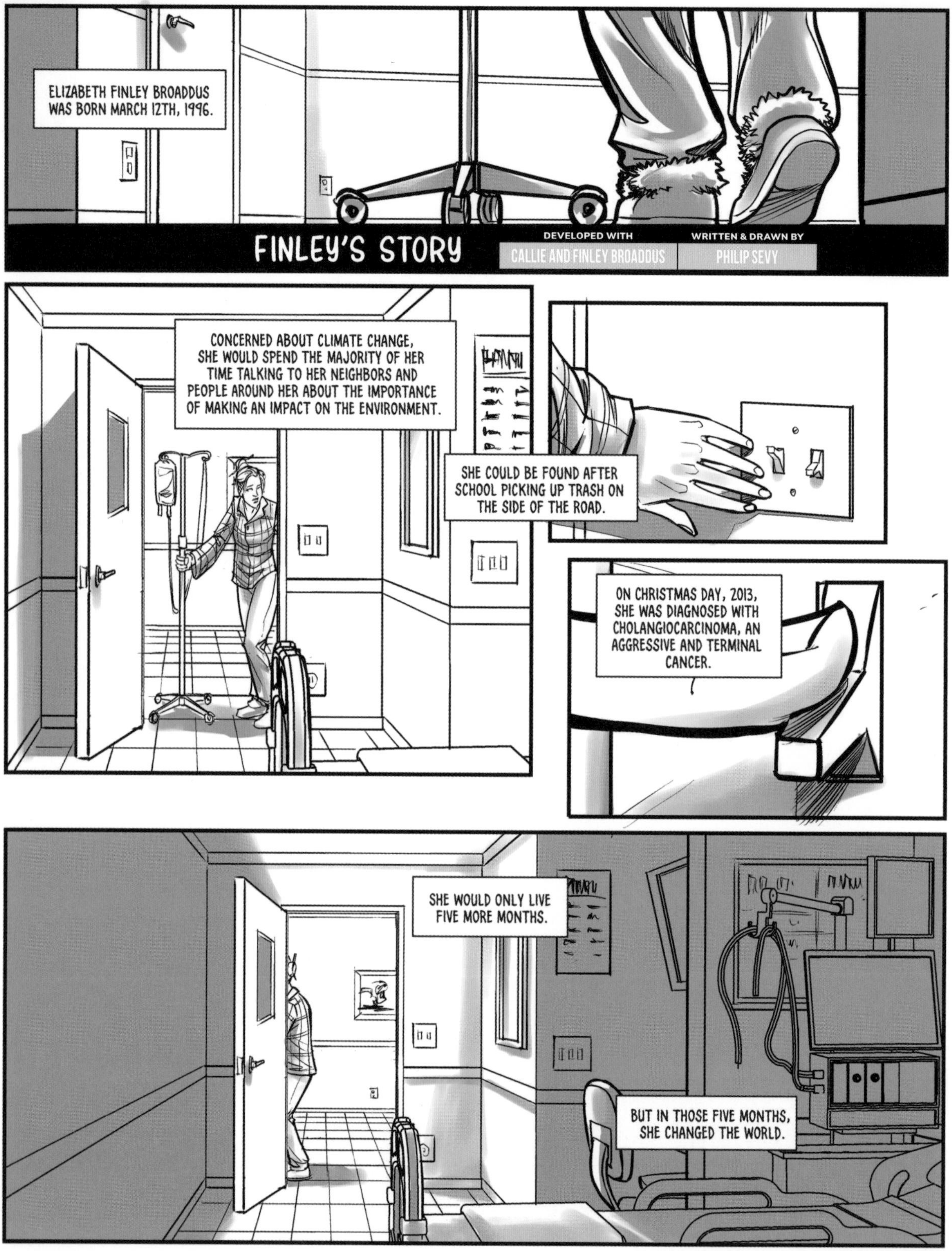

FINLEY'S STORY

DEVELOPED WITH CALLIE AND FINLEY BROADDUS

WRITTEN & DRAWN BY PHILIP SEVY

CONCERNED ABOUT CLIMATE CHANGE, SHE WOULD SPEND THE MAJORITY OF HER TIME TALKING TO HER NEIGHBORS AND PEOPLE AROUND HER ABOUT THE IMPORTANCE OF MAKING AN IMPACT ON THE ENVIRONMENT.

SHE COULD BE FOUND AFTER SCHOOL PICKING UP TRASH ON THE SIDE OF THE ROAD.

ON CHRISTMAS DAY, 2013, SHE WAS DIAGNOSED WITH CHOLANGIOCARCINOMA, AN AGGRESSIVE AND TERMINAL CANCER.

SHE WOULD ONLY LIVE FIVE MORE MONTHS.
BUT IN THOSE FIVE MONTHS, SHE CHANGED THE WORLD.

WHILE IN THE HOSPITAL, SHE STARTED RECEIVING LOTS OF STUFFED ANIMALS AND FLOWER BOUQUETS.
SHE KNEW THAT GIFTS LIKE THESE WERE ONLY CAUSING MORE PROBLEMS FOR THE ENVIRONMENT, SO SHE ASKED THAT PEOPLE PLANT A TREE AND SEND HER A PICTURE OF IT INSTEAD.
WHEN PICTURES CAME POURING IN FROM ALL OVER THE WORLD, SHE KNEW THAT SHE COULD USE HER POSITION TO INSPIRE PEOPLE TO DO MORE.
WITH HER FAMILY, SHE STARTED THE "FINLEY'S GREEN LEAP FORWARD" FUND WITH A GOAL TO RAISE $18,000 BY HER 18TH BIRTHDAY-- A MERE THREE WEEKS AWAY.
ON HER BIRTHDAY, HER FAMILY PRESENTED HER A CHECK FOR THE FULL AMOUNT RAISED ALONG WITH A LIST OF ALL 432 DONORS.
THE CHECK WAS FOR OVER $70,000.

WITH THE MONEY RAISED, FINLEY PICKED THE FIRST TWO GRANTEES--
THE GREEN BELT MOVEMENT IN NAIROBI, KENYA, AND THE CACAPON INSTITUTE IN WEST VIRGINIA, WHICH BOTH FOCUS ON FOREST RESTORATION AND TREE-PLANTING EFFORTS.
FINLEY INTENTIONALLY CHOSE A LOCAL AND A GLOBAL EFFORT, AND SELECTED PROJECTS THAT MAKE A POSITIVE IMPACT ON THE DAILY LIVES OF PEOPLE IN THE SURROUNDING COMMUNITY.
OVER THE YEARS, THE FUND HAS GIVEN OUT GRANTS TOTALING $300,000 AND COUNTING.
THE TREES PLANTED IN HER HONOR CONTINUE TO GROW AND INSPIRE OTHERS TO PLANT MORE AROUND THE WORLD.
FINLEY PASSED AWAY ON JUNE 2ND, 2014.

DURING HER MONTHS OF TREATMENT IN THE JOHNS HOPKINS PEDIATRIC ONCOLOGY WARD, FINLEY WOULD GET UP AND WALK AROUND THE UNIT.

ON MOST DAYS, SHE ONLY HAD THE ENERGY TO DO THIS ONCE.

AS SHE MADE HER ROUNDS, SHE WOULD STOP AT ALL THE EMPTY ROOMS AND TURN OFF THE LIGHTS TO CONSERVE ENERGY.

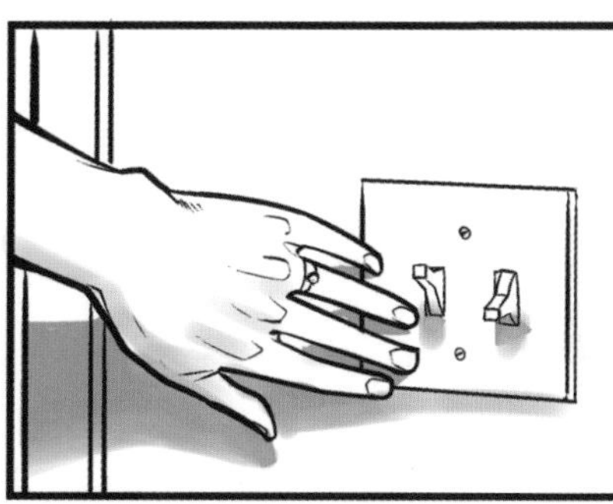

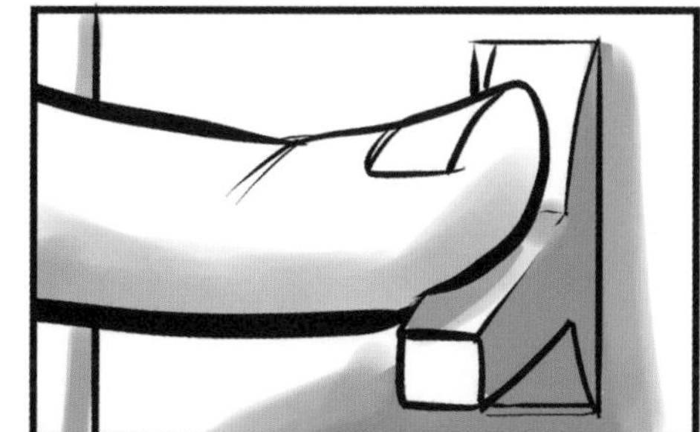

EVEN SEVEN YEARS AFTER HER PASSING, NURSES STILL COMMENT TO FINLEY'S FAMILY THAT THEY STOP IN EMPTY ROOMS AND TURN OFF THE LIGHTS BECAUSE OF THE IMPACT SHE MADE ON THEM.
YOU DON'T CHANGE THE WORLD BY GRAND GESTURES. YOU DO IT THROUGH INSPIRING OTHERS BY THE SMALL ACTIONS YOU TAKE.

SCROLL. REPEAT.
THE MOTH
DEVELOPED WITH BEATIE WOLFE
WRITTEN BY MARIAH MCCOURT
ART & COLORS BY YENNY LAUD
LETTERING BY LUCAS GATTONI
BiNG BiNG BiNG
RECHARGE
DRINK
EAT

escape
crawl
free

One day,
We will Simply be

SPACE
LOST & FOUND
CURIOSITY
Oddity
SLOW DOWN
LOVE
LIGHT
BREATHE
Joy
WHEW!
drip
drip
drip

They waited so long, to put you in a box
But nothing can break your spirit
They tried to change you, until you made them stop
And now you're so high above it

Learn
Evolve
Grow
The End

MY NAME IS SARAH MINTRAM.
I'M MIXED RACE AND WORKING CLASS. FROM LIVERPOOL.
BABY ON BOARD
DEVELOPED WITH SARAH MINTRAM
WRITTEN BY ALEX SEGURA
ART BY STEVEN AUSTIN
COLORS BY PAUL GOODENOUGH
LETTERING BY JIM CAMPBELL
Le Monde world, earth
I DON'T COME FROM A BACKGROUND OF ACTIVISM. WE WERE HAVING A VERY NICE LIFE. ENGLISH TEACHERS IN THE FRENCH COUNTRYSIDE. WE WENT VEGAN. WE WERE BOTH POLITICALLY MOTIVATED.
BUT WE FELT WE WEREN'T DOING ENOUGH.
CAN YOU BELIEVE THIS?
I HAD NO IDEA...
WE BOTH FOUND EXTINCTION REBELLION AT THE SAME TIME. HE SAW A VIDEO ON YOUTUBE, I READ ABOUT IT IN 2018. IT WAS STRANGE WE BOTH SAW IT AT AROUND THE SAME TIME.
TELL THE TRUTH
We are facing an unprecedented global emergency. Life on Earth is in crisis: scientists agree we have entered a period of abrupt climate breakdown, and we are in the midst of a mass extinction of our own making.
Watch part 1 of the Extinction Rebellion talk (above) to see how we are heading for extinction.
THE MESSAGE RESONATED WITH US.
WE HAD TO ACT TO SALVAGE WHAT WAS LEFT OF THIS WORLD.

WE ROSE THROUGH THE RANKS, SERVING "XR" ON AN INTERNATIONAL SCALE--WE GOT CONNECTED, AND IT BECAME A FULL-TIME GIG. FOURTEEN-HOUR DAYS.
MY HUSBAND BECAME A REGIONAL LIAISON. I FOCUSED ON AFRICA AND ASIA.
STOP ECOCIDE
NEW LIVES IN DANGER
YOU LEARN FAST WITH XR. ABOUT THE RISKS. THE CULTURES, AND POLITICAL ISSUES EACH COUNTRY FACES WHEN IT COMES TO THE LOOMING CLIMATE DISASTER.

WILL YOU...?

YES, OF COURSE!

BUT WE HAD A LIFE, TOO. WE GOT ENGAGED. WE TALKED ABOUT CHILDREN, BUT YOUR WORK SEEPS INTO YOUR LIFE.

I...I JUST DON'T KNOW. I LOVE CHILDREN.

I WANT CHILDREN... BUT...

I KNOW. BUT HOW CAN WE...?

DID WE WANT TO BRING A CHILD INTO THIS WORLD? WE WERE TERRIFIED.

AS WE WENT ON, WE LEARNED ABOUT ACTIONS. ABOUT THE RHYTHM OF A PROTEST.

AND THE EXPECTED RISK.

YOU HAVE TO KNOW THE RISKS, AND KNOW YOUR STANCE.

I'M HERE AND I'M NOT MOVING! NO MATTER WHAT! ARREST ME IF YOU WANT!

STOP ECOCIDE

GIVE THE CHILDREN A CHANCE

REPUBLIC

THE EXPERIENCE WAS NEW TO ME.
I'D SEEN POLICE, OF COURSE. BUT I'D NEVER BEEN ON THE OPPOSITE SIDE.
BACK OFF, ALL OF YOU. NOW.

GET THE FRACK OUT!!
THE KIDS TALK DID IT. A TURNING POINT, I GUESS. WE FIGURED IF YOU GOT ARRESTED AT A PROTEST, YOU'D GET A YEAR OR TWO SUSPENDED SENTENCE. THEN WE LEARNED IT'D BE CLOSER TO FIFTEEN **YEARS.**
IT WAS STRANGE TO THINK THAT EVEN THOUGH WE'D ONLY BEEN MARRIED FOR FIFTEEN MONTHS THAT HE COULD BE IN JAIL FOR MORE THAN TEN TIMES THAT.

BIRTH STRIKERS EXIST. I APPLAUD THEM. AND IF WE'RE SERIOUS ABOUT THE CAUSE AND WANT TO HAVE KIDS, WE NEED TO BE SERIOUS ABOUT OUR ACTIONS.

I COME FROM A STRONG IRISH BACKGROUND. FAMILY IS IMPORTANT. I KNEW I WANTED KIDS. REALLY WANTED. IT WAS A HARD CHOICE. BUT COULD I GIVE UP SOMETHING SO IMPORTANT?
WE FELT LIKE WE NEEDED TO WALK THE WALK.

THANK YOU SO MUCH. YOU INSPIRE ME.

YOU MADE A CHOICE FOR ALL OF US.

...FOR SHOWING ME WE CAN BE MOTHERS AND ACTIVISTS...

SO MANY WOMEN REACHED OUT. FROM ALL OVER. SRI LANKA, EVEN. SAYING HOW MUCH THEY APPRECIATED THAT WE DID THIS.

THE STAKES HAVEN'T CHANGED. CLIMATE EMERGENCIES WILL MEAN WARS, FOOD SHORTAGES, CLASS DIVIDES, EXTINCTION EVENTS. IS IT FAIR TO BRING A CHILD INTO THIS WORLD?

CENSORED

BABY ON BOARD

THE ANSWER IS HARD TO DEFINE.

BUT WE WOULD TRY TO CHANGE THE WORLD AS MUCH AS WE COULD, BEFORE AND AFTER OUR CHILD ARRIVED.

BROOKLYN BECKHAM

Photographing animals in the wild was an unforgettable experience. The thought of them suffering, or a future where they don't exist, is unbearable to me.

We all have to fight for wildlife and our planet, and each do all that we can.

HOW YOU CAN HELP

- Avoid places that offer wildlife interactions or shows.
- Pledge not to buy a wild animal as a pet, or products made from wild animal body parts.
- Support local, ethical tours to observe animals safely and responsibly in the wild, where they belong.

With support from World Animal Protection.

Photography by Brooklyn Beckham.

A GLIMPSE
WRITTEN BY
ROBERT KIRKMAN
ART & COLORS BY
CHARLIE ADLARD

CUP
climate
WELCOME
ENJOY THE PERFORMANCE

Rover Red Charlie

WRITTEN BY
GARTH ENNIS
ART & COLORS BY
CHAEL DiPASCALE
LETTERING BY
JIM CAMPBELL

HELLO.
I'M CHARLIE, AND THESE ARE MY FRIENDS RED AND ROVER.

IT FEELS FUNNY TALKING TO FEEDERS AGAIN. IN MY WORLD, YOU'RE ALL GONE.

SOME TIME BACK, WE WALKED FROM THE BIG SPLASH TO THE BIGGER SPLASH.
WE CLIMBED THE BIG BIG AND CROSSED THE BIG EMPTY. WE WERE BOUND FOR THE PLACE WHERE THE SUN LIES DOWN.
HOW'S MY ASS, DUDE?

WE FOUGHT THE HISSPOTS, AND HERMANN THE MONSTER. WE LEARNED TO EAT QUACKERS AND BORK-BORKERS.
WE MET BINGO AND SASHA AND AUDIE AND ALBERT. AND HOBBY, THE SADDEST DOG OF ALL.
NOT NOW, RED.

BUT WE WENT BECAUSE ALL THE FEEDERS HAD GONE. EVERY LAST ONE. IN LESS TIME THAN IT TAKES FOR TWO FEEDINGS.
WE WERE TOLD WE COULD FIND MORE FEEDERS. BUT WHEN WE GOT TO THE BIGGER SPLASH, THERE WERE NONE.
AW, JUST A QUICK SNIFF FOR YOUR OLD PAL.

WHAT WE DIDN'T EXPECT... WAS THAT EVER SINCE THEN... IN A WORLD WITHOUT FEEDERS...
HFF
HFF
WE'VE DONE JUST FINE.
ALL GOOD, RED.

BUT IF *YOU* FEEDERS DON'T CHANGE YOUR WAYS VERY SOON, YOU'RE GOING TO RUIN YOUR WORLD FOR *EVERYONE*.
OUR WORLD IS A DOG PLACE. YOURS WILL BE NO PLACE.
GET IT?

WHO YOU TALKING TO, CHARLIE?
FEEDERS.
I THOUGHT THEY ALL DISAPPEARED.

YEAH, WELL, THEY'RE WORKING ON IT.

The Future in a Plastic Cup
IT WAS THE EARLY 2000s. I WAS STILL A STUDENT IN PHILADELPHIA. A FEW FRIENDS AND I HAD DECIDED TO GO TO A FOLK FESTIVAL.
IT WAS A THREE-DAY AFFAIR, ON FARM-LAND OUT IN THE MIDDLE OF NOWHERE, CAMPED IN TENTS BY THE BANKS OF A RIVER WHOSE NAME I NOW FORGET.

I REMEMBER PICKING OUT A PLACE, PITCHING A TENT IN THE SCORCHING SUN.
THERE WAS NO ALCOHOL PERMITTED BUT WE SNUCK BOTTLES IN ANYWAY.
OTHER MORE ADVENTUROUS SUBSTANCES COULD BE FOUND IF YOU KNEW THE RIGHT PEOPLE.

WE DANCED AND SANG AND REVELED INTO THE NIGHT UNTIL THE SUN SNUCK UP ON US ONCE AGAIN.

WHEN IT GOT TOO HOT, WE WALKED TO THE NEARBY RIVER, STRIPPED OFF AND DOVE IN.

I REMEMBER THAT WATER--CRISP, FULL OF LAUGHTER AND SUNLIGHT.

ON ONE OF THE NIGHTS, SAT BY A ROARING CAMPFIRE, I MET A MAN NAMED SMOKE, WHO SOLD TIE-DYE T-SHIRTS AND JEWELRY MADE OF BONE.

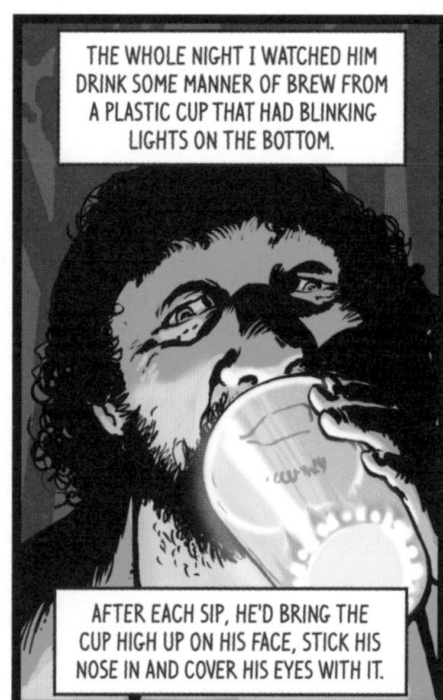
THE WHOLE NIGHT I WATCHED HIM DRINK SOME MANNER OF BREW FROM A PLASTIC CUP THAT HAD BLINKING LIGHTS ON THE BOTTOM.
AFTER EACH SIP, HE'D BRING THE CUP HIGH UP ON HIS FACE, STICK HIS NOSE IN AND COVER HIS EYES WITH IT.

WHAT'S IN THE CUP, SMOKE?
I ASKED HIM.
THE FUTURE, MAAAAAN.
HE SAID.

WE ALL LAUGHED.

LATER, AS WE WERE LEAVING THE FARM TO GO BACK HOME, I FOUND THE CUP ON THE GROUND, ABANDONED.

I LOOKED IN AND WONDERED WHAT MANNER OF FUTURE LAY AT THE BOTTOM OF A PLASTIC CUP.
THEY STOPPED THE FESTIVAL SOMETIME AROUND 2013. I STILL FIND MYSELF THINKING ABOUT THOSE DAYS OF MUSIC AND FRIENDSHIP. I HOPE THAT CAREFREE PLACE STILL EXISTS.
I HOPE YOU CAN STILL JUST DIVE INTO THE RIVER AND LOOK AT YOUR OWN FEET THROUGH THE CLEAR WATER BELOW.
THESE PLACES OF JOY AND ABANDON. PLACES WHERE YOU CAN REACH OUT AND TOUCH THE WORLD AND KNOW THAT IT TOO IS **ALIVE**, JUST LIKE YOU.
WE'RE LOSING THESE PLACES. SO MANY OF THEM HAVE ALREADY BEEN RELEGATED TO MEMORIES.
I DON'T IMAGINE THEY EXIST IN THE FUTURES TO BE FOUND AT THE BOTTOM OF PLASTIC CUPS.
WRITTEN BY RAM V
ART & COLORS BY SI FRASER
LETTERING BY FERNANDO FUENTES MÉNDEZ
END

NEW YEAR'S EVE, 2012.
NO DOUBT THE PREVIOUS GENERATIONS LEFT US IN A HOLE.

NEW YEAR'S EVE, 2022.
NO SENSE WASTING ANY ENERGY ON WHO'S TO BLAME. IT'S OUR INHERITANCE.

NEW YEAR'S EVE, 2032.
IT'S EASY TO FEEL HOPELESS.

NEW YEAR'S EVE, 2042.
...THAT'S BECAUSE THINGS LOOK HOPELESS.

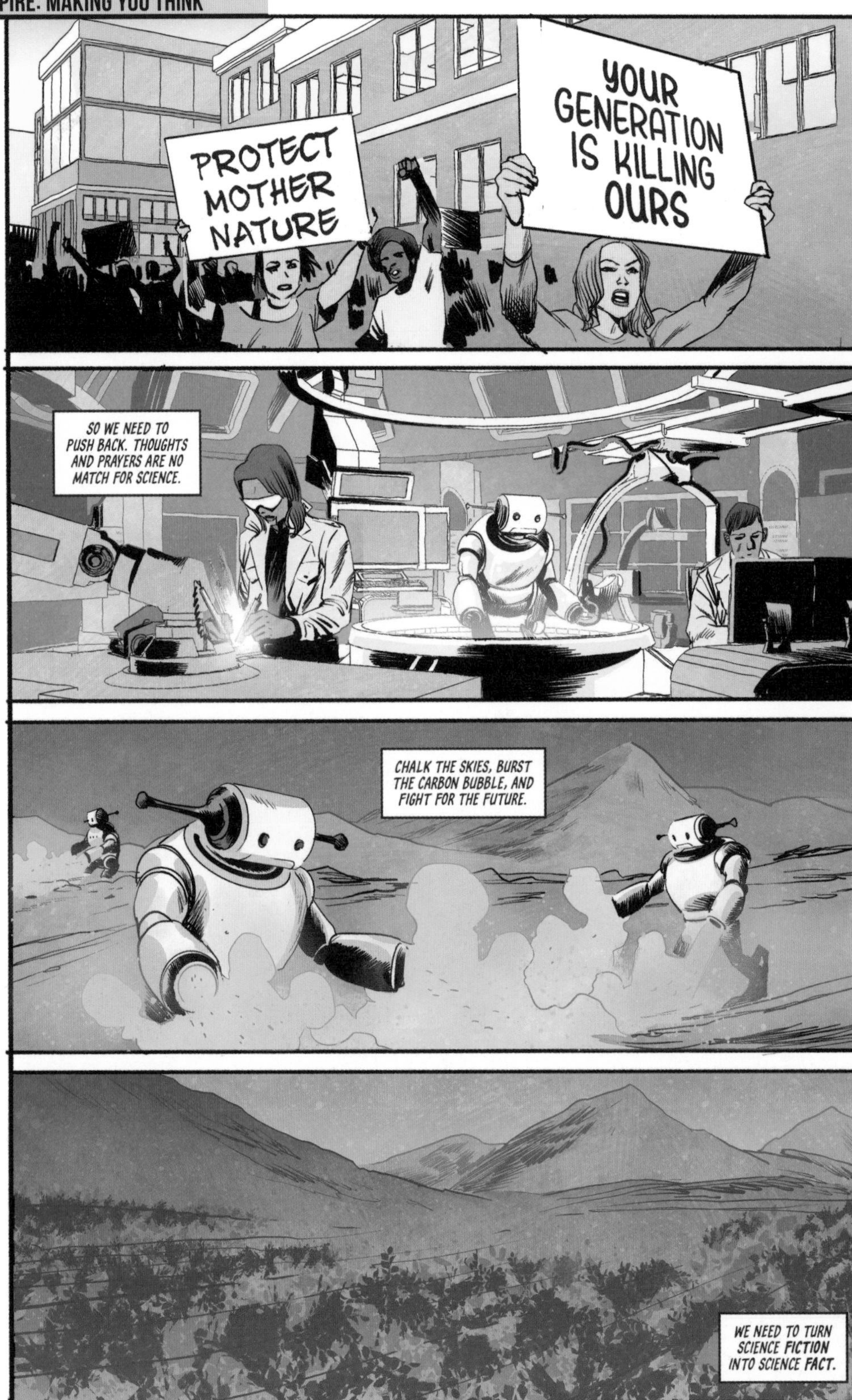
PROTECT MOTHER NATURE
YOUR GENERATION IS KILLING OURS
SO WE NEED TO PUSH BACK. THOUGHTS AND PRAYERS ARE NO MATCH FOR SCIENCE.
CHALK THE SKIES, BURST THE CARBON BUBBLE, AND FIGHT FOR THE FUTURE.
WE NEED TO TURN SCIENCE FICTION INTO SCIENCE FACT.

IT IS 2052.
CAUSE WE'RE THE ONLY HELP THAT IS COMING.
TURNED AROUND
WRITTEN BY
GERRY DUGGAN
ART BY
RAFAEL ALBUQUERQUE & RAFAEL PIMENTEL
COLORS BY
NICK FILARDI
LETTERING BY
BERNARDO BRICE

I HATE THIS PLACE.
ARE YOU AN IDIOT? WHAT WERE YOU THINKING!
I WAS STOPPING TO AVOID AN ANIMAL! YOU WENT RIGHT INTO ME!
EVERYTHING IS ANGRY. COMPRESSED. CONSTRICTED.
YOU CAN'T BREATHE HERE.
THE AIR IS THICK WITH SMOKE. THE GRASS HAS A SICKLY COLOR TO IT.
GIANT CITY BLOCKS BLOT OUT THE SKY. WE ONLY SEE THE SUN FOR AN HOUR OR SO BEFORE IT HIDES BEHIND THEM.
AND STILL THEY BUILD MORE. BECAUSE WHO NEEDS THE SUN?
I FEAR FOR THE YOUNG. THEY'LL INHERIT THIS MESS ONE DAY.
THEY WANTED US TO GO TO A FARM! IN THE CITY! WITH ANIMALS AND EVERYTHING!
I MADE MUM WRITE ME A NOTE TO GET OUT OF IT.
THEY'LL NEVER KNOW THE JOY OF AN OPEN, GRASSY FIELD.
A SECLUDED WOODLAND IN THE SUMMER.
BIG ISSUE? NO? SPARE SOME CHANGE?
PLEASE, I NEED TWENTY QUID FOR A NIGHT IN THE HOSTEL.
ALL THAT'S LEFT ARE SCRAPS OF LAND WHERE WE CAN TRULY FEEL OURSELVES--
--WHILE WE'RE EVICTED FROM OUR HOMES.
I SAID GIVE IT A REST!
HE STARTED IT! IDIOT STOPPED IN THE MIDDLE OF THE ROAD!
I WAS AVOIDING A FOX!
SURVIVAL OF THE FITTEST DOESN'T COUNT ANYMORE. NOBODY'S GOING TO SURVIVE.

SOON TO BE AN EXCITING APARTMENT COMPLEX
NO GRASS
THE WHOLE WORLD WILL BE CONCRETED OVER, AND WE'LL SUFFOCATE.
THEY DON'T UNDERSTAND, YOU SEE.
WE NEED THESE PARKS, THESE OPEN SPACES MORE THAN MONEY MADE FROM NEW HOUSES.
IT'S A DISGRACE. THEY'RE NOTHING BUT ANIMALS.
THE OLD LADY'S RIGHT. THERE'S NO CONNECTION WITH NATURE ANYMORE. PEOPLE STARE AT GLOWING SCREENS RATHER THAN SUNSETS.
COME ON, LOVE. NO BEGGING HERE. MOVE ON.
METROPOLITAN POLICE
MOVE WHERE? THEY EVICTED ME FROM MY HOUSE!
I'VE GOT NOWHERE!
BUT SUNSETS LEAD TO DARKNESS.
AND THE DARKNESS IS COLD, WHEN YOU'RE LEFT OUTSIDE.
I SAID GET OFF ME!
WE JUST CONSTRICT IN PLACE, STRUGGLING TO BREATHE AS THE CITIES CLOSE IN AROUND US--
--UNTIL WE BREAK.
CLATTER
THE STREET GOES SILENT. YOUNG AND OLD, RICH AND POOR, FED AND STARVING-- THE GULFS BETWEEN ARE TOO WIDE FOR THEM.
THINGS WILL GET WORSE BEFORE THEY GET BETTER. PEOPLE NEED TO HAVE A COLLECTIVE MOMENT OF UNITY, TO REMEMBER THAT THEY WERE ONCE A TRIBE--

--AND SO I REVEAL MYSELF.
VERMIN. MANGE-RIDDLED SCAVENGERS. SHOULD BE EXTERMINATED. WE'VE HEARD THEM ALL.
BUT THROUGH ALL OF THIS, WE SURVIVE. WE FIND THE SMALL PATCHES OF NATURE IN YOUR CONCRETE PRISONS AND LIVE.
AND WHEN WE ARE SEEN, IT'S MAGICAL. A GOOD OMEN. A MOMENT TO STOP, AND CONSIDER.
A MOMENT TO REGAIN SANITY.
THERE REALLY WAS A FOX! GOOD ON YOU FOR NOT HITTING THE POOR THING!
WE DON'T SPEAK MUCH, BUT IF WE DID? WE'D TELL YOU TO STOP DESTROYING YOUR GREEN PLACES.
I'LL SORT YOU A PLACE FOR TONIGHT, OKAY?
TO BE THANKFUL FOR WHAT YOU HAVE, AND TO HELP THOSE THAT HAVE NOT.
NOT SO FANTASTIC, MR. FOX
DEVELOPED WITH JUDGE ROB RINDER
WRITTEN BY TONY LEE
ART BY NEIL MCCLEMENTS
LETTERING BY JIM CAMPBELL
NOT JUST TO KEEP THE AIR FRESHER, BUT TO HELP YOUR OWN MENTAL HEALTH.
AND WE'D REMIND YOU THAT YOU ARE A PACK. A TRIBE.
SO ACT LIKE ONE.
LISTEN TO FOXES. BECAUSE WE'RE LISTENING TO YOU. BUILD MORE GREEN SPACES IN YOUR CONCRETE CITY PRISONS--
--AND WE MIGHT JUST COME VISIT YOU MORE.

Come on, Mama! I want to see the sharks... And bats!! And pangolin!!
Honey, please. You know I don't like going there.
MMMAAAAMMMAAA!!!
DEVELOPED WITH
MALAIKA VAZ
WRITTEN BY
PAUL GOODENOUGH
ART & LETTERING BY
ROSITSA VANGELOVA
WOOOOWWW!!!
Were they REALLY alive when you were a girl?
END WILDLIFE MARKETS FOREVER

WRITTEN AND DRAWN BY
POLYP

WRITTEN BY	ART BY	LETTERING BY
ALAN MOORE	MELINDA GEBBIE	BERNARDO BRICE

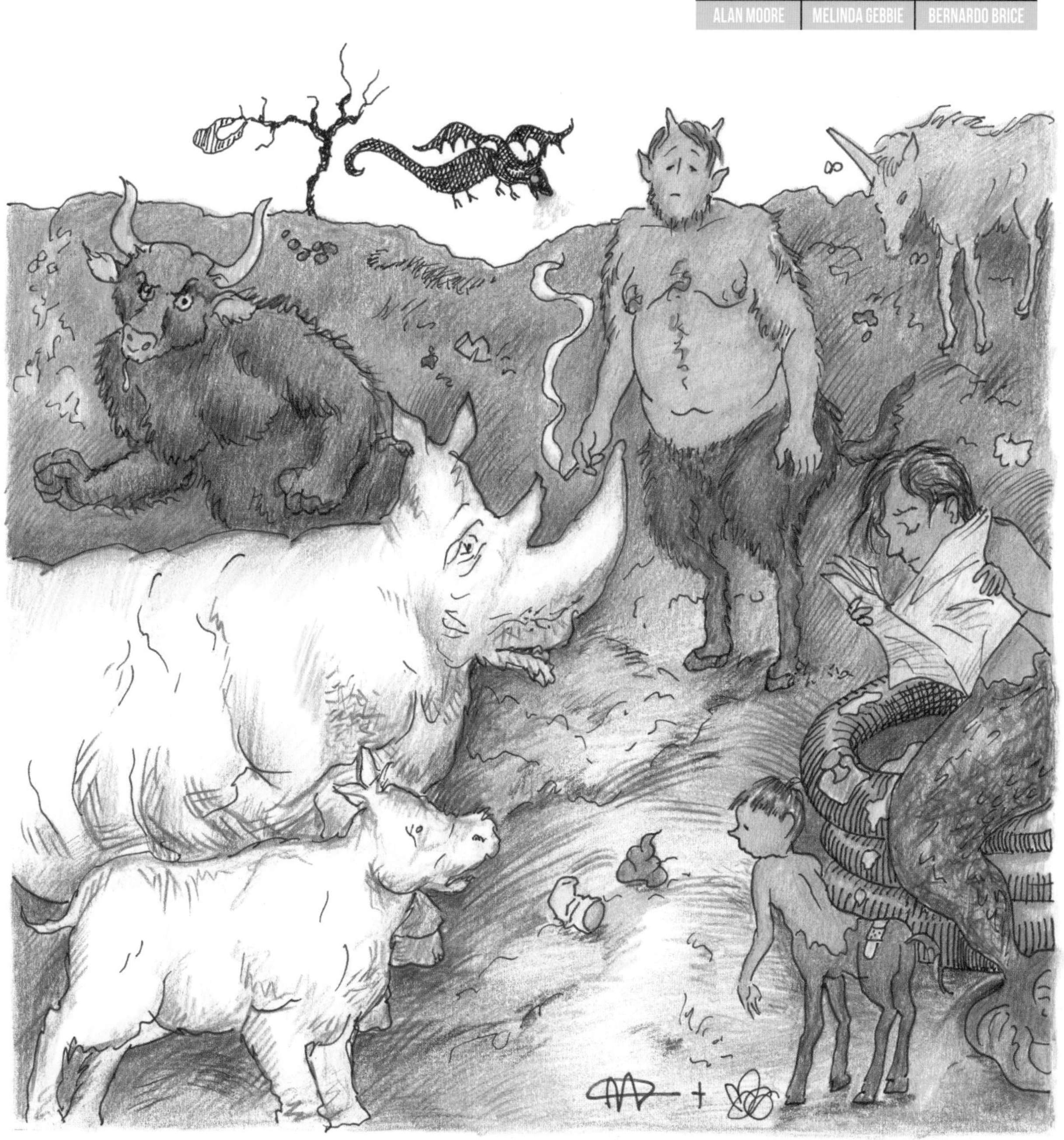

"Frankly, Trevor, this isn't looking good."

WRITTEN AND DRAWN BY
JOEL PETT

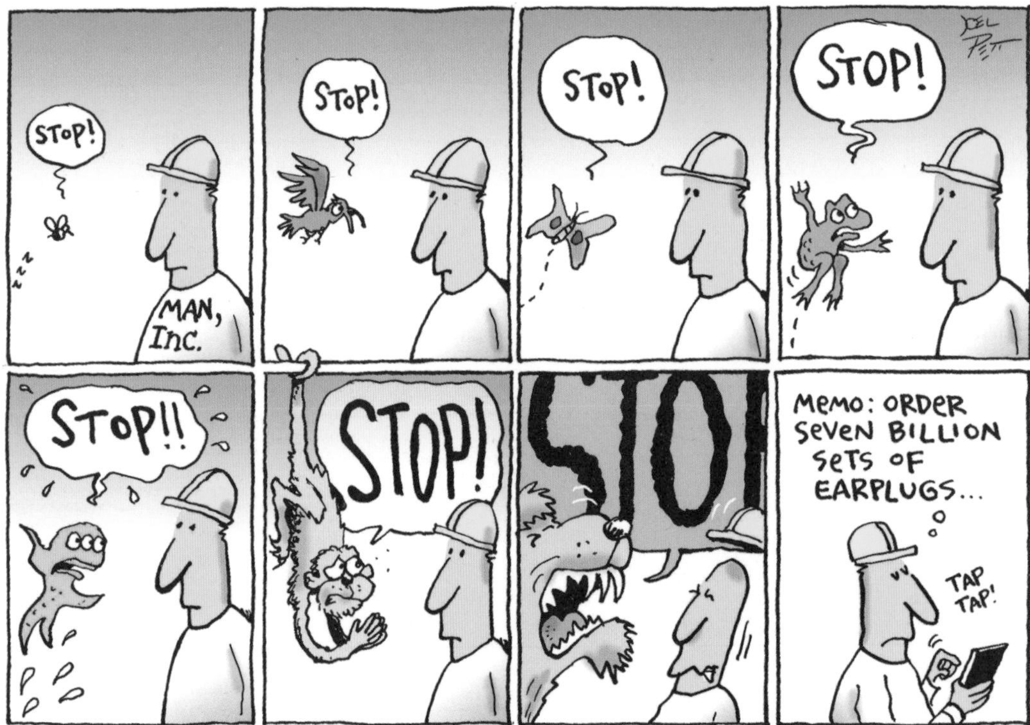

THE RETURN OF
PEDANTIC STAN
THE COMICS FAN
IN:
CLIMATIC CLIMATE CRISIS!
COMIC STRIP BY
JOHN FREEMAN & LEW STRINGER

IN AN UNKEMPT, FORGOTTEN GRAVEYARD, SOMETHING STIRS...
NO, I'M SORRY, BUT I JUST CANNOT ALLOW MYSELF TO REMAIN DEAD AND NOT COMMENT ON THE CURRENT STATE OF THE COMICS INDUSTRY!
STAN FLAN 19.. - 1991 KILLED BY COMICS

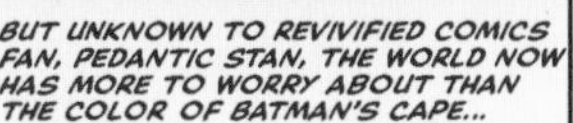

BUT UNKNOWN TO REVIVIFIED COMICS FAN, PEDANTIC STAN, THE WORLD NOW HAS MORE TO WORRY ABOUT THAN THE COLOR OF BATMAN'S CAPE...

ALL THESE COMICS... WHERE DO I START READING? I DON'T RECOGNISE ANY OF MY FAVOURITES!
OLD SCHOOL, HUH? YOU CAN READ THEM ON THIS...
AREN'T YOU WORRIED ABOUT THE ZOMBIE IN THE SHOP?
I'M MORE WORRIED WE'RE GOING TO BE FLOODED BEFORE WE SHUT!
URG!
ARG!
OW!
OOF!
COMIX
COMIC COMIC
EEK!
SERIOUS COMICS
CRISIS ON EARTH NOW

THIS DEVICE IS A COMPUTER? WHERE'S THE MOUSE?
YOU DON'T NEED ONE OF THOSE! WHERE HAVE YOU BEEN FOR THIRTY YEARS?
CONTEMPLATING THE MISSING HALF PAGE IN A MARVEL UK WARHEADS COMIC... WHAT HAPPENED TO IT?

NO PLASTIC BAGS TO PRESERVE DEAD PAPER... YOU COULD HAVE YOUR ENTIRE COMIC COLLECTION ON THIS!
WAIT, I'VE BEEN DEAD SEVENTEEN YEARS, THREE MONTHS AND TWO DAYS--I WONDER WHAT MUM'S DONE WITH ALL MY COMICS?
ERM... ENVIRONMENTALLY FRIENDLY LANDFILL?

NOOOOO!
FLIPPING ZOMBIES! NO CLIMATE CRISIS TRAINING!

WHAT'S HAPPENED TO THE WORLD? ELECTRIC CARS?
WHY AREN'T PEOPLE USING PLASTIC BAGS FOR THEIR SHOPPING?
HAS THE WORLD GONE MAD?

MUM! I'M HOME!
EEEEK!
PLEASE TELL ME MY ROOM IS AS I LEFT IT SEVENTEEN YEARS, THREE MONTHS AND TWO DAYS AGO!

STANLEY! DON'T YOU DARE DECOMPOSE ON MY STAIR CARPET!

MY COMIC COLLECTION! WHAT HAVE YOU DONE, MUM?
I SOLD IT OFF ON EBAY THREE YEARS AGO, LOVE. HAD A LOVELY HOLIDAY IN THE SOLOMON ISLANDS, BEFORE THEY VANISH...

I THINK WE ALL HAVE A LOT MORE TO WORRY ABOUT THAN A FEW MOULDY OLD COMICS NOW, SON.
STOP THE FUTURE! I WANT TO GET OFF!
YOU AND ME BOTH, BUD!

PEDANTIC STAN'S PARTICULARLY PEDANTIC PETULANT PANEL. NUMBER ONE OF ONE. CUT OUT AND KEEP.
HMM, READERS, DID YOU KNOW BUYING A NEW MOBILE DEVICE CONSUMES AS MUCH ENERGY AS USING AN EXISTING PHONE FOR AN ENTIRE DECADE? * IT SEEMS MY INNATE DESIRE TO HOLD ONTO MY LOVELY PAPER COMICS MIGHT BE A BETTER IDEA, EH? EH?
WHEN DID YOU LAST WASH?
I'M EMBRACING MY INNER ECO-NESS.
EEEEW!
* JOURNAL OF CLEANER PRODUCTION VOLUME 177, 10 MARCH 2018, PAGES 448 - 463.

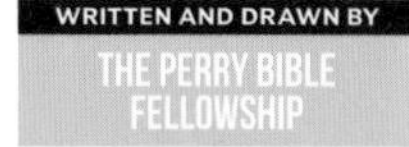

The Talk

Amends

URTH-01

WRITTEN AND DRAWN BY MICHAEL LEE HARRIS

LETTERING BY FRANK CVETKOVIC

GULA!
GULA!

WHAT IS IT, ACERVUS?! THE COMMS ARE FOR OFFICIAL USE ONLY!

THIS IS OFFICIAL! IT'S SO COLD HERE I CAN SEE MY BREATH!

SO?!
"SO"?! YOU'RE IN CHARGE OF ENVIRONMENTAL MAINTENANCE!
Food Storage
OH RIGHT!

YEAH, YEAH, YEAH, I'M ALREADY THERE.
OFFLINE
OFFLINE

FIXING IT AS WE SPEAK...
SCHLUUURP!
SCHLURP SCHLURP

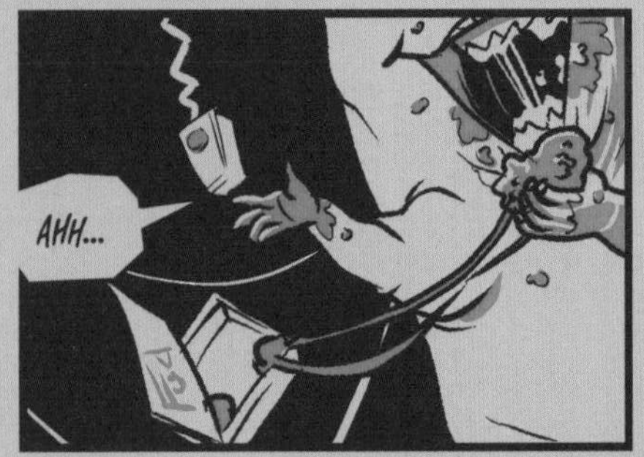
AHH...

FZZT
NOT THE LIGHTS AGAIN!

FWEPP

FZZT
LUUUDIOOOOO!

FIX THE POWER!
RAWR
YEAH, YEAH, YEAH. I'M IN THE POWER ROOM RIGHT NOW...

Power: 0.02%
FIXING IT AS WE SPEAK...

DO YOU KNOW WHEN IT WILL BE FINISHED?!
PEW PEW
UM... I JUST NEED SOME PARTS, UM... FABRICATED...

ACERVUS?!
WHAT?!

I NEED SOME STEM BOLTS FABRICATED TO STABILIZE THE POWER. WHEN CAN YOU GET THOSE?

UM...
NINJA CHEERLEADERS!
NINJA CHEERLEADERS!
NINJA CHEERLEADERS!
REE
REE

Fabricator
87% Finish
NINJA
EERLEADERS

Ship Total
Raw Material
Remaining:
0.04%

UH...

VMMMM

ACERVUS?! WHAT'D YOU DO?

IT WASN'T ME! I'M NOT THE ONE HOGGING ALL THE POWER, **LUDIO!**
WHAT?!
YOU'RE IN CHARGE OF THE FABRICATORS!
AND **GULA!** WHY AREN'T ENVIRONMENTAL CONTROLS WORKING?!

UM... BECAUSE... UM...
UM... **THE FABRICATORS!**
WE'RE GONNA FREEZE TO DEATH AND IT'S ALL **YOUR FAULT!**

WELL, AT LEAST WE STILL HAVE FOOD...
WE **DO** STILL HAVE FOOD... RIGHT, GULA?
STAY AWAY FROM ME!
THIS IS ALL **YOUR FAULT!**
NO! IT'S YOURS!
I'M GONNA KILL YOU!
IT'S BOTH YOUR FAULTS!

LEN'S ANIMAL MAGIC

WRITTEN BY	ART BY	COLORS BY	LETTERING BY
SIR LENNY HENRY	MARK BUCKINGHAM	LEE LOUGHRIDGE	TODD KLEIN

Z
O
O
I USED TO WORK AT DUDLEY ZOO WHEN I WAS 14-ISH.

I DID ALL SORTS. WASHED DISHES. MOPPED FLOORS.
THEY EVEN LET ME SERVE SOMETIMES.

THE BOSS TRIED TO MAKE IT LOOK LIKE I WAS THE RIGHT AGE...
BAR

SOMETIMES ON MY LUNCH BREAK I'D HAVE A QUICK STROLL ROUND THE ZOO. I LOVED LOOKING AT ALL THE ANIMALS...
THE HYENAS WERE PROBABLY MY FIRST AUDIENCE.

MY WIFE'S A RED HEAD.
NO HAIR, JUST A RED HEAD.

MY DOG CAN TALK.
I SAID: HOW'D YOU FEEL THIS MORNING?
HE SAID: RUF.

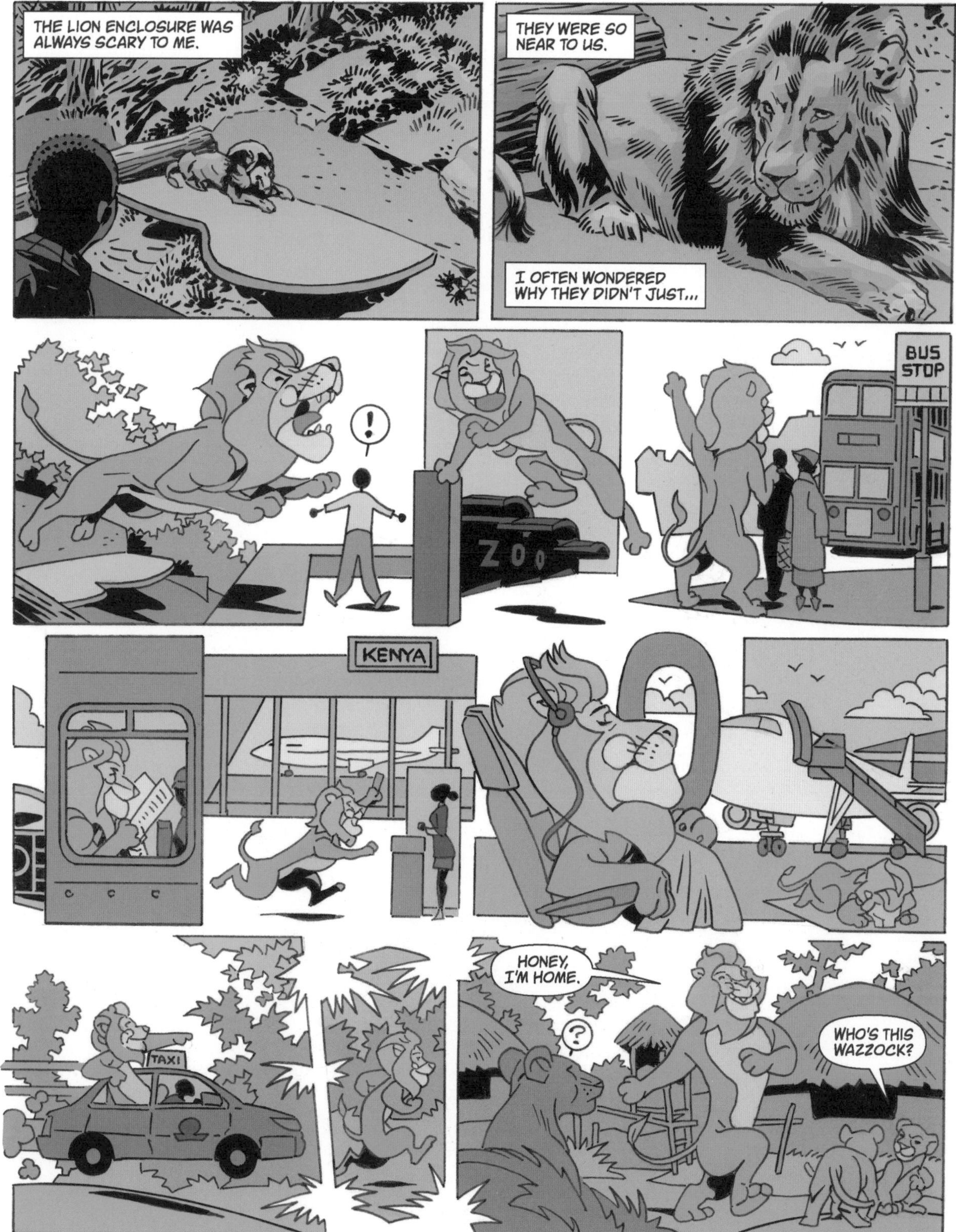
THE LION ENCLOSURE WAS ALWAYS SCARY TO ME.
THEY WERE SO NEAR TO US.
I OFTEN WONDERED WHY THEY DIDN'T JUST...
!
ZOO
BUS STOP
KENYA
TAXI
HONEY, I'M HOME.
?
WHO'S THIS WAZZOCK?

I WAS ALWAYS THINKING OF THINGS LIKE THAT WHEN I WASN'T WORKING.
I GUESS THAT'S WHERE ALL THE STUPID STUFF COMES FROM.

DUCKS USED TO SCARE ME.
THE DRAKES WERE ALWAYS VERY ROUGH ON THE MALLARDS-- LITERALLY CHASE, CHASE, CHASE-- AND THEN BITE THE NECK AND HAVE THEIR WAY WITH THEM IN BROAD DAYLIGHT.

I ALWAYS THOUGHT IT WAS UNJUST.
MADE ME WANT TO GO INTO...

...DUCK LAW.
YOUR HONOUR, THE CCTV CAMERA TELLS US THIS REPREHENSIBLE MALLARD TOOK THIS DUCK BY FORCE. NO DINNER, NO MOVIE, NO NOTHING.
YOUR HONOR, I OBJECT! THAT COULD HAVE BEEN ANY DUCK DOING WHAT HE DID AND SHE KNOWS IT.
THIS FOOTAGE IS THE TRUTH OF WHAT HAPPENED THAT DAY.
THE TRUTH? YOU CAN'T HANDLE THE TRUTH!

I ALWAYS THOUGHT CUDDLES THE KILLER WHALE NEEDED A BIGGER POOL. AT LEAST THREE FOOTBALL PITCHES' WORTH. I THINK HE WAS CONSTRAINED BY HIS SURROUNDINGS.

AT LEAST THAT'S WHAT IT FELT LIKE TO ME.

I USED TO GO TO THE DISCO AT THE TOP OF DUDLEY ZOO. THE QUEEN MARY BALLROOM.

YOU HAD TO WALK PAST THE ELEPHANT ENCLOSURE TO GET THERE.

I ALWAYS THOUGHT IT WAS WEIRD FOR THE ELEPHANTS WATCHING US GO IN AND THEN SEEING US COME OUT LATER THAT NIGHT, IN A RIGHT STATE.

I WONDER WHAT THEY THOUGHT OF US?

EVERY WEEK WE SEE THIS. THEY ENTER SMARTLY CLOTHED, CLEAN, SMELLING FRESH.
NO TROUBLE AT ALL.
I KNOW.

AND THEN THEY COME OUT OF THAT PLACE. DRUNK.
FIGHTING LIKE DOGS.
ROLLING AROUND IN THE BUSHES, FORNICATING.
AND WE'RE THE ONES THEY BLOODY LOCK UP.

HAPPY DAYS THOUGH. NOT SURE THE ANIMALS FELT THE SAME WAY. BUT FOR ME? THEY WERE HAPPY DAYS...
ZOO
END

Mommy?
It's time to go.
Are we going upstairs?
No.
Why not? You said you'd take us.
I said I would *try.*
Can we try now?
There's no more trying. Let's wake your sister.
Honey, wake up.
Daddy?
This is Mommy.
She always forgets.
Let's not poke fun.
Well, she never learns.
She's younger than you. She'll be the one having the last laugh.
Come on, both of you. It's time to go.
Upstairs?
No, to the *room.*
We're not allowed to go to the room.
We can do what we want now.

It's so bright.

It has to be. Light keeps them alive.
They're alive?

Of course.

Sleeping?
Yes.
Are they dangerous?
Not like this.

A boy and a girl?
Yes.

So they have friends?
Yes.

There's more?
You have everything.
Just enough to start over.
We need to keep them safe.
Horses!
Zebras.
And those, look at their necks.
Giraffes. You've never seen one.
Are they like dogs?
That one will be as tall as five of you.
When will it get to grow?
You'll like this room. It's my favorite.
Soon, I hope.

Wow!
Do you like it?
I think it's my favorite too.
Try to get comfy.
Where's yours?
These are for you.
Why are you *sad,* mommy?
No, no, not sad. People cry when they're *happy* too.
When you wake up, you can try again.
But for now...
...sleep.
NOA
WRITTEN BY
SCOTT "BABYDADDY" HOFFMAN
ART BY
JUAN BOBILLO
LETTERING BY
BERNARDO BRICE

HE IS FOURTEEN WHEN HE CLIMBS THE THOUSAND BUND.

HE HAD LISTENED THE NIGHT BEFORE, HEARD THE SOUND OF THE GIANT TIDES RECEDING, THEIR FETID STENCH MUTED, AND DECIDED IT WAS TIME.

HERO

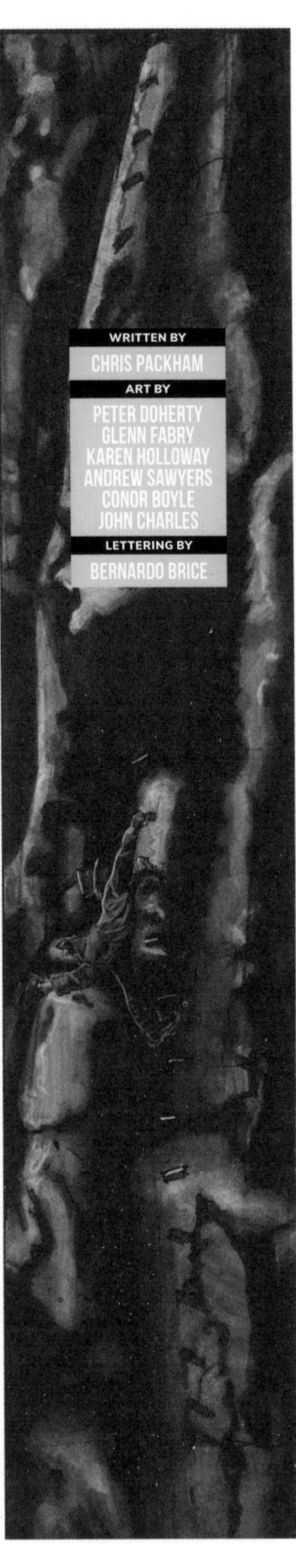
WRITTEN BY
CHRIS PACKHAM
ART BY
PETER DOHERTY
GLENN FABRY
KAREN HOLLOWAY
ANDREW SAWYERS
CONOR BOYLE
JOHN CHARLES
LETTERING BY
BERNARDO BRICE

HE IS TIRED OF LIVING IN THE SHADOWS, BENEATH THE GIANT WALL THAT KEEPS THE MASSIVE SEAS AT BAY--

--TIRED OF SCRAPING A MISERABLE EXISTENCE AMONG THE RUINS OF WHAT USED TO BE.

THERE'S A WIDE WORLD OUT THERE--CREATURES THE LIKE OF WHICH THEY CAN ONLY DREAM NOW.

HE'S SEEN THEM. SEEN THEM ON THE SKIN OF THE **LIBRARY OF LIFE.**

SHE'D READ BOOKS, SEEN PICTURES. SHE WAS A LIVING RECORD.
WHEN I DIE THEY'RE GOIN' TO SKIN ME, BOY, HANG IT UP FOR SHOW.
ONCE HE'D SEEN A BURD, JUST LIKE ONE ON THE LIBRARY OF LIFE. OR THOUGHT HE HAD --
BURD!

BITTA OLD PLASTI.
THINGS AREN'T LIKE THEY USED TO BE.

YOU DREAMERS GET THAT THOUGHT OUT OF YOUR DAMN HEADS! THE BURDS ARE GONE! IT'S ALL GONE!

YEARS BEFORE, NOT LONG AFTER HIS FATHER AND BRUD DIED IN THE THIRD FAMINE, HIS MOTHER TOOK HIM TO SEE THE TREE.

THE LINE OF PILGRIMS WAS SO LONG--THEY'D TURNED BACK WITH JUST A GLIMPSE OF ITS SAD, TWISTED SILHOUETTE.
THE WORLD'S LAST TREE, THEY SAID.
BUT THEY DIDN'T KNOW EVERYTHING.

THERE WERE WONDROUS THINGS OUT THERE, BEYOND THE BUND, THINGS THEY'D HEARD RUMORS OF--GLOWING NEON CITIES, WHERE THE LUCK-BORN LIVED.

AND BEYOND THEM...WHAT? MORE. THERE MUST BE MORE.
IT CAN'T ALL BE DEAD AND TWISTED AND ROTTING AWAY...

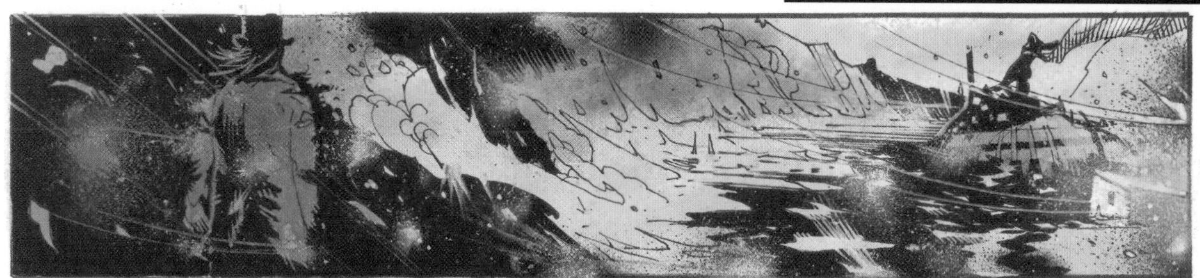

WHAT YOU DOIN'?
EHHH...?

FISHING. THAT'S WHAT I'M DOIN'. I'M THE FISHERING MAN.
FISH--IS THAT LIKE A WHALE?
WHAT DO YOU KNOW ABOUT WHALES?
I SEEN 'EM. ON THE LIBRARY OF LIFE. SHE GOT ALL SORTS ON HER, THINGS THAT USED TO BE.

WELL, THERE'S NO WHALES. THERE'S NO FISH. WE KILLED 'EM ALL.
US?
PEOPLE.

SO WHY'RE YOU FISHERIN'?

COS WE GOTTA KEEP TRYIN'. KEEP LOOKIN'. YOU GOTTA HAVE HOPE, BOY.

SO I KEEP TRYIN', AND THAT'S ALL I CATCH. CARBAGE. NOTHIN' ELSE LEFT.
THERE'S RATS.
YEAH, THERE'S RATS. AN' BUGS. AN' US. RATS AN' BUGS AN' US.

DON'T KNOW WHICH IS THE BIGGEST PEST.
'LEAST YOU CAN EAT RATS AN' BUGS. GUESS THAT MAKES IT US.
NOW CLEAR OFF, BOY. GOT A FISH TO CATCH.

HE CLIMBS THE THOUSAND BUND OFTEN AFTER THAT, AND ALWAYS THE FISHERING MAN IS THERE.

UNTIL ONE DAY HE'S NOT.
'COURSE THEY'LL ALL BE GONE SOON ENOUGH. IF THE NEXT FAMINE DOESN'T GET THEM, THE NEXT PLAGUE WILL.

NOBODY LIVES LONG ANYMORE. THEIR TEETH ROT AND THEIR SKIN FLAKES AND THEIR NAILS FALL OFF. POISONED, LIKE THE LAND.
WHAT HAPPENED TO HIM?
I DON'T KNOW. BIG WAVE MAYBE. SWEPT AWAY.

YOU GOING TO COME BACK?
IF I GET TIME. GONNA BE BUSY. YOU KNOW WHERE I'LL BE.

SO CLIMB THAT THOUSAND BUND, AND THAT'S WHERE YOU'LL SEE HIM--

ANY TIME OF DAY, ALL KINDS OF WEATHER, HE'LL BE THERE, THE FISHERING MAN.

BECAUSE SOMEBODY'S GOT TO DO IT. **SOMEBODY'S** GOT TO KEEP HOPE ALIVE.
FISH?!

SINCE I WAS A CHILD I WAS FASCINATED BY NATURE.

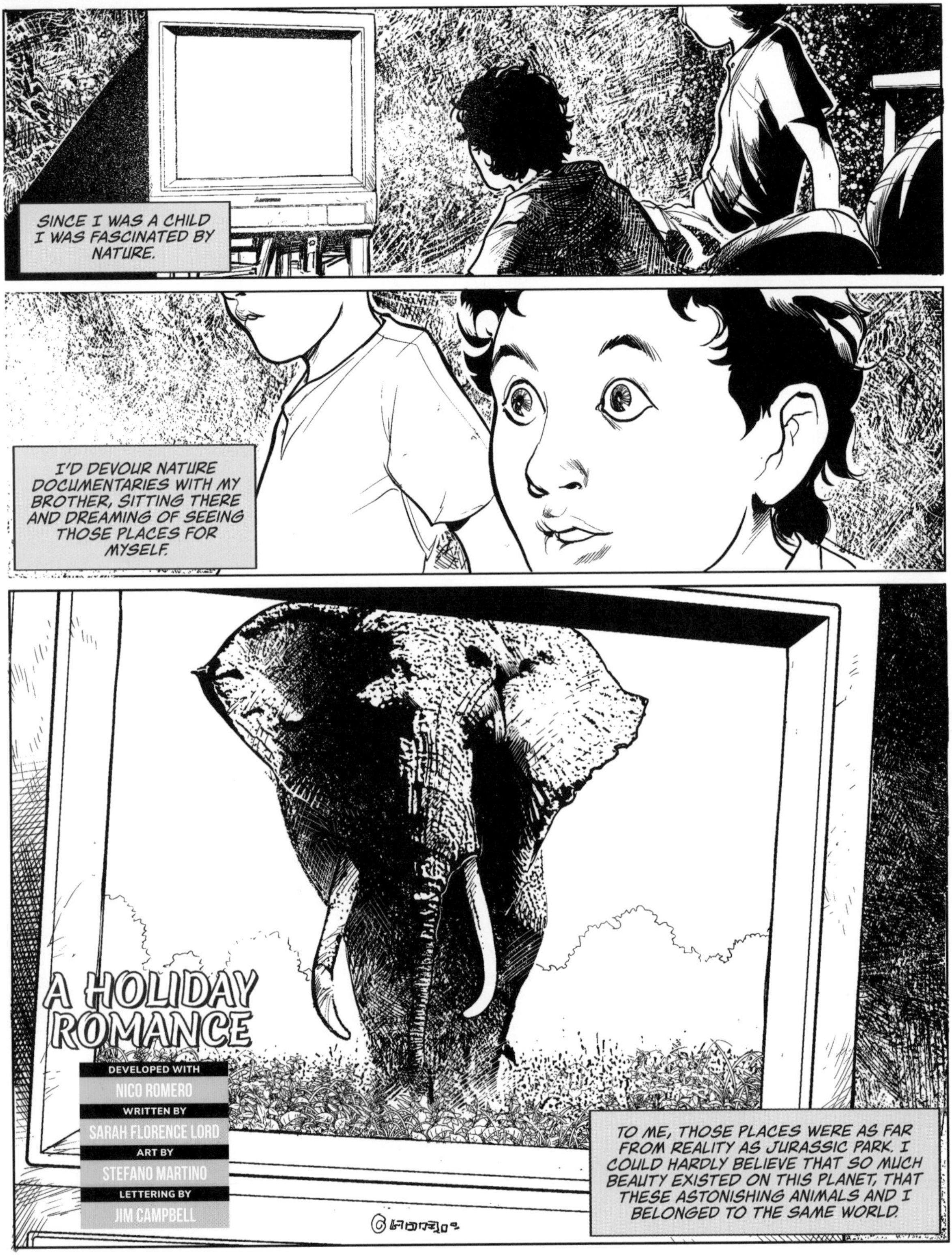

I'D DEVOUR NATURE DOCUMENTARIES WITH MY BROTHER, SITTING THERE AND DREAMING OF SEEING THOSE PLACES FOR MYSELF.

A HOLIDAY ROMANCE

DEVELOPED WITH
NICO ROMERO
WRITTEN BY
SARAH FLORENCE LORD
ART BY
STEFANO MARTINO
LETTERING BY
JIM CAMPBELL

TO ME, THOSE PLACES WERE AS FAR FROM REALITY AS JURASSIC PARK. I COULD HARDLY BELIEVE THAT SO MUCH BEAUTY EXISTED ON THIS PLANET, THAT THESE ASTONISHING ANIMALS AND I BELONGED TO THE SAME WORLD.

IT WASN'T UNTIL I GREW UP AND MADE THE TRIP TO UGANDA FOR MYSELF THAT I THINK I REALLY BELIEVED THOSE NATURE DOCUMENTARIES WERE REAL.

IT WAS SO DIFFERENT FROM THE WORLD I WAS USED TO, BUT I HAD NEVER FELT MORE MYSELF, MORE AT HOME...

I SAW THINGS THAT MADE ME LAUGH...

...FRIGHTENED ME...

...AND MADE ME CRY. WATCHING THE GORILLAS IS SOMETHING I'LL NEVER FORGET. TO ME IT WAS NATURE IN ITS PUREST FORM.

BEFORE I LEFT SPAIN, I WAS ANXIOUS ABOUT LEAVING BEHIND EVERYTHING THAT CONNECTED ME TO MY REALITY. BUT THAT WAS BEFORE I REALIZED THIS WAS MY REALITY, AND THOSE DISCOMFORTS WERE ACTUALLY FREEDOMS.

BUT EVERY ADVENTURE HAS TO COME TO AN END.

WE ALL HAVE TO COME BACK TO REALITY EVENTUALLY.

BUT IT DOES MAKE YOU WONDER...
Metro
Manuel Becerra

DOES IT REALLY HAVE TO BE THAT WAY?

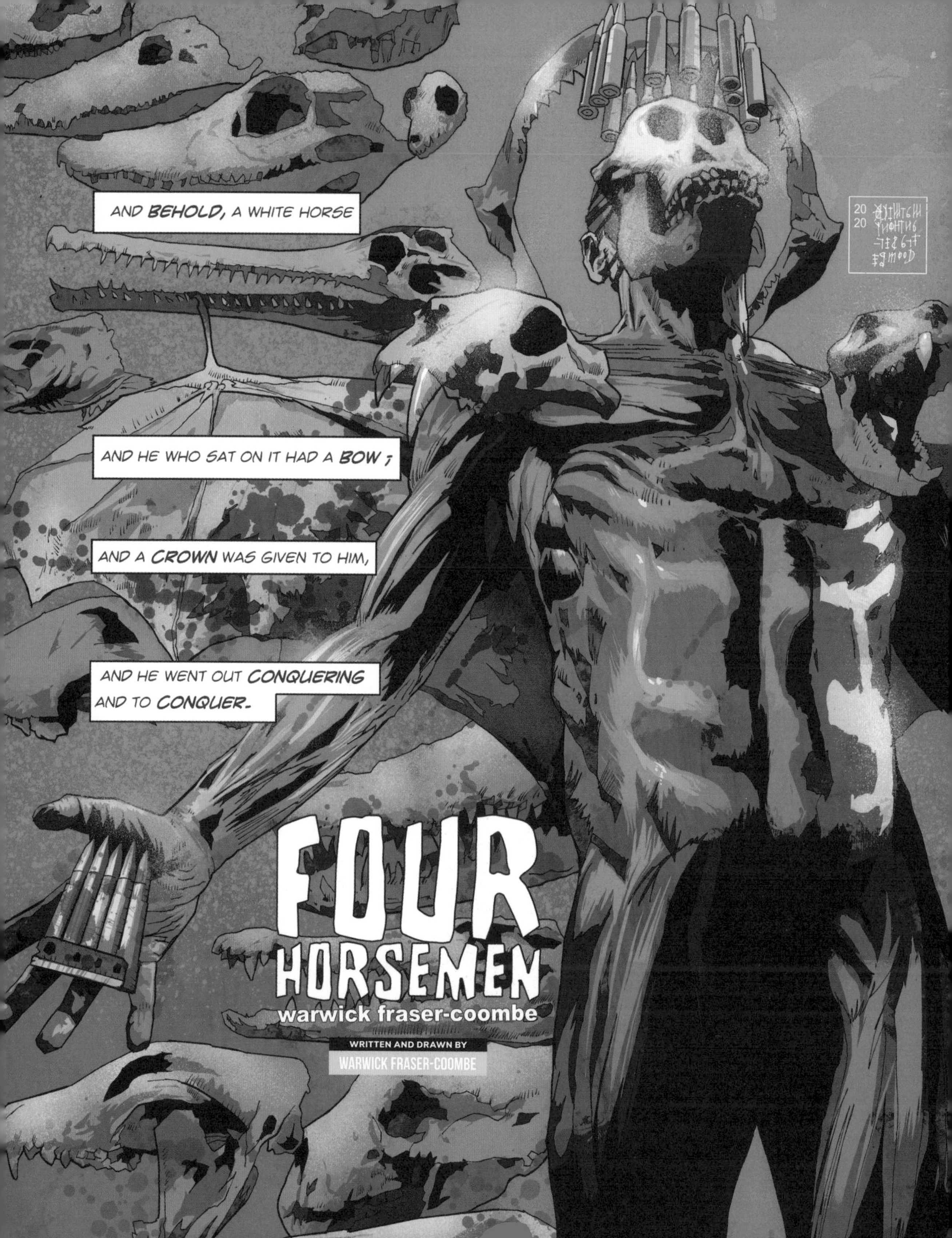
AND BEHOLD, A WHITE HORSE
AND HE WHO SAT ON IT HAD A BOW ;
AND A CROWN WAS GIVEN TO HIM,
AND HE WENT OUT CONQUERING AND TO CONQUER.
FOUR HORSEMEN
warwick fraser-coombe
WRITTEN AND DRAWN BY
WARWICK FRASER-COOMBE

"COME." AND ANOTHER, A RED HORSE, WENT OUT;
AND TO HIM WHO SAT ON IT, WAS GRANTED TO TAKE PEACE FROM THE EARTH,
AND A GREAT SWORD WAS GIVEN TO HIM.

"COME-"

I LOOKED, AND **BEHOLD,**

A **BLACK HORSE**

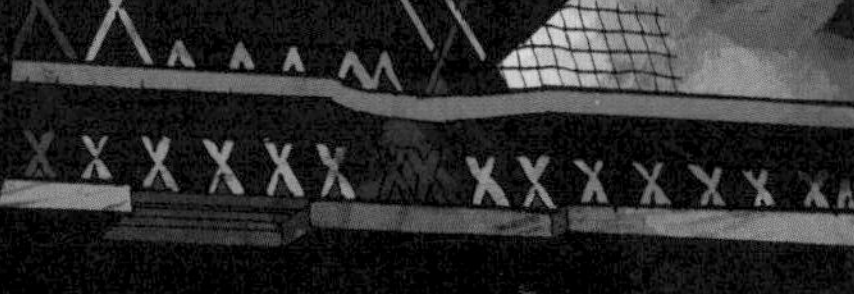

"COME-"
I LOOKED, AND BEHOLD,
A PALE HORSE
AND HE WHO SAT ON IT HAD THE NAME DEATH
AND HADES WAS FOLLOWING WITH HIM

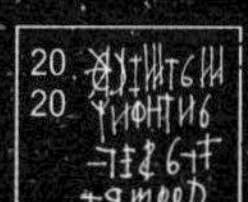

SAVE THE THINGS YOU **LOVE!**

WITH THANKS

STORY CREDITS

p.12 **Chapter 1 - Change The System** © Dan Shearn.
pp.13-15 **Make Earth Great Again** © Paul Goodenough, Serg Acuña, and Bernardo Brice.
p.16 **Relationships Can Change** © Paul Goodenough and Sarah Graley. With Tom Mustill.
pp.17-20 **Melody** © Kieron Gillen and Sean Phillips. With Cara Delevingne and EcoResolution.
p.21 **The Bunker** © Paul Goodenough and Dami Lee. With Cara Delevingne and EcoResolution.
p.22 **Smother Earth** © Hannah Hillam. With Jojo Mehta.
p.23 **Info on Ecocide** © Jojo Mehta.
p.24 **Ecocide** © Paul Goodenough and Hannah Hillam. With Jojo Mehta.
pp.25-31 **The XR Diaries** © John Wagner, Leonardo Marcello Grassi, and Bernardo Brice. With Liam Norton, Fazeela Mubarak, Cathy Eastburn, and Detective Sergeant Paul Stephens.
pp.32-33 **Man on the Moon** © Bella Lack, Morgan Beem, and Jim Campbell.
pp. 34-39 **The Concrete Ceiling** © Paul Goodenough and Rosita Vangelova. With Luisa Neubauer and Fridays for Future.
pp.40-45 **Roots & Shoots** © Chuck Austen, Paul Goodenough, Lee Carter, and Jim Campbell. With Roots & Shoots and Jane Goodall.
pp. 46-50 **Dima's Story** © Pornsak Pichetshote, Peter Gross, JP Bove, and Bernardo Brice. With Dima Litvinov.
p.51 **Meat Free Monday** © Lunarbaboon (Chris Grady). With Paul McCartney, Mary McCartney, Stella McCartney, and Meat Free Monday.
pp. 52-54 **Own the Skies** © Rob Williams, Ben Oliver, and Bernardo Brice. With George Monbiot.
p.55 **Consumed to Death** © George Monbiot and Ian Stopforth.
p.56 **How Much Did That Cost: Make-Up** © Dave Schneider, Amber Weedon, and War and Peas (Elizabeth Pich and Jonathan Kunz). With Cheddar Gorgeous.
p.57 **How Much Did That Cost: Phones** © Drew Sheneman. With Prof. Ben Garrod.
p.58 **How Much Did That Cost: Chocolate** © Paul Gooodenough and Daniella Orsini. With Prof. Ben Garrod.
p.59 **Prawn Free** © Jamie Woolley, Davey Jones, and Bernardo Brice. With Anton Litvanov.
p.60 **Aware Animals: Away** © Aware Animals (Mira Petrova).
p.61 **Aware Animals: Repercussions** © Aware Animals (Mira Petrova).
p.62 **Purchasing Power** © Paul Goodenough and Hector Trunnec. With Michelle Desilets and Helen Buckland.
p.63 **Eat Healthy Eat Wise** © Paul Goodenough and Moksha Carambiah.
p.64 **Is the Price Right?** © Jamie Woolley and Stilly. With Will Travers OBE.
p.65 **How Much Is Land *Really* Worth?** © Paul Goodenough, and Stilly. With Will Travers OBE and Diane Coyle.
p.66 **Dye Another Day** © Lucy Siegle, Paul Goodenough, and goodbadcomics (Aditi Mali).
p.67 **Not All Jeans Go to Heaven** © Sarah Florence Lord, Paul Goodenough, and Things in Squares. With Candiani.
p.68 **Cruelty What** © War and Peas (Elizabeth Pich and Jonathan Kunz).
p.69 **Fabulous** © War and Peas (Elizabeth Pich and Jonathan Kunz).
pp.70-71 **Humanity: Annual Report** © Peter Milligan and Guillermo Ortego.
pp.72-75 **In a Yellow Wood** © Michael Conrad, Noah Bradley, Lee Loughridge, and Bernardo Brice. With John Sauven.
pp.76-79 **Just Business** © Amy Chu, E.J. Su, and Bernardo Brice. With Steve Backshall.
p.80 **Fast Fashion Kills** © Paul Goodenough, Patrick Goddard, Dylan Teague, and Jim Campbell.
p.81 **Is Fast Fashion Making Us Sick?** © Moksha Carambiah. With The Plastic Soup Foundation.
pp.82-84 **Do You Know What Your Money Is Funding?** © Lucky Generals, Doug Braithwaite, Sebastian Cheng, and Bernardo Brice.
p.85 **Pension Power!** © Make My Money Matter.
p.86 **While You Sleep** © Paul Goodenough and Abhijeet Kini. With Richard Curtis and Victor Solís.
p.87 **Woke** © War and Peas (Elizabeth Pich and Jonathan Kunz). With Richard Curtis.
pp.88-89 **In My Town, There Was a Hole...** © Cheddar Gorgeous, Paul Goodenough, and John McNamee.
pp.90-92 **Marshall Islands** © Sarah Florence Lord, Matylda-Mai McCormack-Sharp, and Bernardo Brice. With Doreen deBrum and Kathy Jetnil-Kijiner.
p.93 **TED Talks** © Paul Goodenough and Sarah Graley. With Peter Gabriel, The Interspecies Internet, and TED.
p.94 **Jyoti Ma** © Sarah Florence Lord, Tyler Jennes, Zoe Thorogood, and Bernardo Brice. With Jyoti Ma and Mindahi Bastida.
pp.95-98 **Lessons from the Lakota** © Sarah Florence Lord, Zoe Thorogood, and Bernardo Brice. With Moses Brings Plenty.
p.99 **Mitakuye Oyasin–We Are All Related** © Native Daily Network.
pp.100-102 **Beneath the Trees** © David Baillie and Juni Ba. With Elizabeth Wathuti.
pp.103-105 **Arthur's Story** © Greg Lockard, Diigii Daguana, and Jim Campbell. With Arthur Golong.
pp.106-108 **Depths** © Corey Brotherson, Geraldo Borges, Lee Loughridge, and Bernardo Brice. With Meena Rajput.
p.109 **Community and Culture** © Meena Rajput.
pp.110-111 **A Climate of Hope** © Roger Stern, Joe Orsak, and Jim Campbell. With Prof. Michael Mann.
pp. 112-113 **Secure Global Ocean Protection** © Greenpeace
p.114 **Chapter 2 - Protect The World** © Dan Shearn.
pp. 115-117 **The Shepherd & The Thunder** © Brian Azzarello, Danijel Zezelj, and Bernardo Brice. With Peter Hammarstedt.
pp.118-121 **The Surfing Hippos of Loange** © Erika Lewis, Alti Firmansyah, Yenny Laud, and Bernardo Brice. With Peter Hammarstedt.
pp.122-123 **Saikou's Nightmare and a Dream** © Abdulkareem Baba Aminu, Ryan Cody, and Bernardo Brice. With Mustapha Manneh and Sainey Gibba.
pp.124-127 **Rainbow Warriors** © Joe Pokaski, Sebastián Píriz, Lee Loughridge, and Bernardo Brice. With Hettie Geenen.
p.128 **Osos polares** © Victor Solís.
p.129 **The Selfless Giant** © Paul Goodenough, Danijel Zezelj, and Bernardo Brice. With Andy Serkis.
p.130 **School's Out** © War and Peas (Elizabeth Pich and Jonathan Kunz).
p.131 **Going Fast!** © Taika Waititi, Paul Goodenough, and Safely Endangered (Chris McCoy).
pp.132-133 **Ohno** © Brian Azzarello and Cliff Chiang.
p.134 **Important Takeaway** © Steve Backshall and Harry Venning.
p.135 **Protect the Seas** © Peter Hammarstedt and Steve White.
p.136 **Tinder Date** © David Schneider, Amber Weedon, and Safely Endangered (Chris McCoy).
pp.137-139 **Jenji** © Simon Furman and Timo Wuertz. With Jonathan Barnard.
p.140 **Deforestation** © World Land Trust, Reserva: The Youth Land Trust, and Neil "Blackbird" Sims.
pp.141-143 **There's Blood on Your Hands** © Tyler Jennes, Aneke Murillenem, Jim Campbell. With Callie Broaddus and Javier Robayo.
p.144 **Barb and Rudy** © The Perry Bible Fellowship (Nicholas Gurewitch).
pp.145-147 **Loss of Our Land** © Ariela Kristantina. With Bradley Hiller.
pp.148-149 **Forest Hero** © Desiree Alvarado and David Baillie. With Paulo Adario.
pp.150-151 **The Universal Language** © KT Tunstall, Tyler Jennes, Craig Cermak, and Bernardo Brice. With Adam Gardner, James Valentine, Stefan Lessard, Diana Ríos, and Nicolás Saba.
pp.152-154 **May the Best Endangered Species Win!** © Tyler Jennes and Sheila Putri. With Callie Broaddus.
p.155 **Grandma's Amazing Adventures** © Paul Goodenough and Momo & Popo. With Malaika Vaz.
p.156 **Just Shoot Us** © Paul Goodenough and Ken Catalino.
p.157 **Bullfight** © Ricky Gervais and Rob Steen.
p.158 **There's No One Like Me** © Sad Animal Facts (Brooke Barker).
pp.159-168 **Mama Bear** © Jenny Jinya and Paul Goodenough.
p.169 **Earth Day** © Dami Lee.

pp.170-173 **Sanctuary** © David Baillie, Conor Doyle, and Paul Goodenough. With Prof. Ben Garrod, Jim Desmond, and Jenny Desmond.
p.174 **Oh, Wise Tree** © War and Peas (Elizabeth Pich and Jonathan Kunz).
p.175 **I Promised** © Dinos and Comics (James Stewart and K Roméy).
pp.176-178 **Bridge to the Future** © Ruth Fletcher Gage, Christos Gage, Mike Perkins, and Bernardo Brice.
pp.179-183 **Born Free Forever** © Tab Murphy, Mwelwa Tax, and Jim Campbell. With Virginia McKenna OBE, David Manoa, Dr Caroline Ng'weno, Raabia Hawa, and Bella Lack.
p.184 **Hedgehog Heroes!** © Hugh Warwick and Stig. With Michael Rosen.
p.185 **What If God Was One of Us?** © Alicia Souza. With Eric Bazilian.
p.185 **Shared Experiences** © Just Comics (Joan Chan). With Paul McCartney, Mary McCartney, and Stella McCartney, and Meat Free Monday.
p.186 **Never Forget** © Paul Goodenough and Buddy Gator (Chow Hon Lam). With Brooklyn Beckham and World Animal Protection.
p.187 **End Wildlife Crime** © John Scanlon and Mister Hope.
pp.188-189 **Keep Wildlife in the Wild** © Born Free
pp.190-196 **Little Mo** © Jenny Jinya.
p.197 **The Nightmare** © Buddy Gator (Chow Hon Lam).
p.198 **Torn** © Amy Meek. With Kids Against Plastic.
pp.199-202 **The Power of Youth** © Alex Segura, George Kambadais, and Jim Campbell. With Aditya Mukarji.
p.203 **30 Million** © Moksha Carambiah.
pp. 204-205 Guatemala's Laguna Grande Reserve © World Land Trust
p.206 **Chapter 3 - Restore the Damage** © Jeff Langevin.
p.207 **Exploring My Options** © Extra Fabulous Comics (Zach Stafford). With Karrueche Tran.
pp.208-209 **The Smallest Seed** © Sarah Florence Lord, Abby Howard, and Bernando Brice. With Pungky Nanda Pratama.
pp.210-212 **Yellow Bird** © Tate Brombal, Ray Fawkes, Jeff Lemire, and Bernando Brice.
p.213 **Rewild the Galapagos** © Ian Stopforth and Paula Castaño.
pp. 214-215 **Rewild Our World!** © Re:wild
pp.216-217 **The Cloud Forest** © Marguerite Bennett, Maia Kobabe, and Bernardo Brice. With Callie Broaddus.
pp. 218-219 **A Future for the Chocó** © Reserva: The Youth Land Trust
pp.220-223 **The Wild Place** © Rik Worth, Jordan Collver, and Owen Watts. With Rebecca Wrigley.
pp.224-228 **Parley** © Si Spurrier, Dani, and Bernardo Brice. With Rewilding Europe.
p.229 **Dark Souls** © Karl Kerschl.
p.229 **Tree Hugger** © Karl Kerschl.
pp. 230-231 **Make Europe a Wilder Place** © Rewilding Europe
pp.232-235 **The Long View** © George Mann, Tazio Bettin, and Nicola Righi. With Sir Charlie Burrell, Issy Tree, Russ Carrington, and Rebecca Wrigley.
pp.236-237 **Better the Devil** © Sarah Florence Lord, Ian Stopforth, and Paul Goodenough. With Dacre Montgomery and Re:Wild.
pp.238-241 **Acceleration** © Corey Brotherson, Kennedy Garza, and Lucas Gattoni. With Alejandro Agag and Extreme E.
pp.242-247 **The Great British Lie** © Paul Goodenough, Simon Myers, and Jim Campbell. With the Wildlife Trusts, Dame Judi Dench, Bella Lack, James McVey, Mya-Rose Craig, Liz Bunnin, Gel Spellman, Dominic Monaghan, Megan McCubbin, Chris Packham, Sir Ian McKellen, Sir Patrick Stewart, Monty Don, Jerome Flynn, Gillian Burke, Prof. Ben Garrod, Tanya Burr, Judge Rob Rinder, Cara Delevingne, and Craig Bennett.
pp. 248-249 **Bring Beavers Back** © The Wildlife Trusts
pp. 250-253 **Regeneration Change** © Chris Ryall, Mike Collins, Candice Han, Sebastian Cheng, and Jim Campbell. With Gabe Brown, and Arizona Muse.
pp.254-257 **A Fevered Mind and a Pot of Pee** © Lucy Lawless, Paul Goodenough, Tyler Jennes, and David Mack.
pp.258-259 **Grow Your Own, Save the World** © Jamie Woolley, Goran Gligovic, and Bernardo Brice. With Prof. David Goulson.
pp.260-267 **Untidy** © Paul Goodenough, Geoff Senior, and Jim Campbell. With Dominic Monaghan.

P.268 **Chapter 4 - Inspire and Educate** © Jock.
pp.269-271 **I Love You, Earth** © Timo Wuertz. With Yoko Ono.
p.272 **Use Your Voice** © Marian Greaves and Dan Bradbury. With Letters to the Earth.
p.273 **What Did You Do During the Climate Crisis?** © Gillian Burke, Tom Mustill, and Guillermo Ortego.
p.274 **Interview With Extinction Rebellion Co-Founder Gail Bradbrook** © John Wagner and Abdulkareem Baba Aminu. With Gail Bradbrook.
p.275 **ClimateTBD** © Rosemary Mosco.
p.276 **The Children's Fire** © Jerome Flynn and David Smart Knight.
p.277 **Rewild For Victory** © Tom Mustill and Maggie Behling.
p.278 **Ask Not Only What You Could Do for Your Countryside** © Tom Mustill and Maggie Behling.
p.279 **Future Archaeologist** © Tom Mustill and Aware Animals (Mira Petrova).
pp.280-283 **Finley's Story** © Philip Sevy. With Callie Broaddus and Finley Broaddus.
pp.284-289 **The Moth** © Mariah McCourt, Yenny Yaud, and Lucas Gattoni. With Beatie Wolfe.
pp.290-293 **Baby on Board** © Alex Segura, Steven Austin, Paul Goodenough, and Jim Campbell. With Sarah Mintram.
p.294 **A World Without Color** © Brooklyn Beckham. With World Animal Protection.
pp.295-300 **A Glimpse** © Robert Kirkman and Charlie Adlard.
p.301 **Red Rover Charlie** © Garth Ennis, Michael DiPascale, and Jim Campbell.
pp.302-304 **The Future in a Plastic Cup** © Ram V, Si Fraser, and Fernando Fuentes Méndez.
pp.305-307 **Turned Around** © Gerry Duggan, Rafael Albuquerque, and Rafael Pimentel, Nick Filardi, and Bernardo Brice.
pp.308-310 **Not So Fantastic Mr Fox** © Tony Lee, Neil McClements, and Jim Campbell. With Judge Rob Rinder.
p.311 **End Wildlife Markets Forever** © Paul Goodenough and Rositsa Vangelova. With Malaika Vaz.
p.312 **Polyp Strips** © Polyp (www.polyp.org.uk).
p.313 **Frankly Trevor** © Alan Moore and Melinda Gebbie.
p.314 **Joel Pett Strips** © Joel Pett
p.315 **The Return of Pedantic Stan, The Comics Fan in: Climatic Climate Crisis!** © John Freeman and Lew Stringer.
p.316 **The Talk** © The Perry Bible Fellowship (Nicholas Gurewitch)
p.316 **Amends** © The Perry Bible Fellowship (Nicholas Gurewitch).
pp.317-321 **Urth-01** © Michael Lee Harris and Frank Cvetkovic.
pp.322-325 **Len's Animal Magic** © Mark Buckingham, Lee Loughridge, and Todd Klein. With Sir Lenny Henry.
pp.326-329 **Noa** © Juan Bobillo and Bernardo Brice. With Scott "Babydaddy" Hoffman.
pp.330-335 **Beyond The Shadows** © Chris Packham, Peter Doherty, Glenn Fabry, Karen Holloway, Andrew Sawyers, Conor Boyle, John Charles, and Bernardo Brice.
pp.336-338 **A Holiday Romance** © Sarah Florence Lord, Stefano Martino, and Jim Campbell. With Nico Romero.
pp.339-343 **Four Horsemen** © Warwick Fraser-Coombe.

INDEX OF CONTRIBUTORS

INDEX OF CONTRIBUTORS...CONT

ACKNOWLEDGMENTS

Works referenced:

p.54 Jean-Jacques Rousseau, non-fiction book, Discourse on Inequality
p.90 Kathy Ketñil-Kijiner, poem, Midnight
p.102 Wangari Maathai, speech at Goldman Awards, 24 April 2006
pp.110-111 Professor Michael E.Mann, non-fiction book, The New Climate War, copyright © 2001. Reprinted by permission of PublicAffairs., an imprint of Hachette Book Group, Inc.
p.145 Sir Nicholas Stern, opinion piece, "The View" in "The South China Morning Post"
p.152 Sir Andrew Lloyd Webber and Sir Trevor Nunn, song, "Memory" from "Cats"
p.183 George Adamson, autobiography, My Pride and Joy
p.185 Eric Bazilian, song, One of Us. Used by Permission / All Rights Reserved
p.267 George Monbiot, non-fiction book, Feral
pp.269-271 Yoko Ono, song, I Love You, Earth. Copyright © Yoko Ono 1985 Used by Permission/ All Rights Reserved. Also, Letters to the Earth campaign
p.272 Marian Greaves, letter, Use Your Voice. Letters to the Earth campaign, which invites letters to be written worldwide in response to the planetary crisis. www.letterstotheearth.com

Photography credits:

The publisher would like to thank the following for their kind permission to reproduce their photographs:
Key: a-above; b-below/bottom; c-center; f-far; l-left; r-right; t-top

Greenpeace: © Greenpeace 6tr, Oscar Siagian / Greenpeace 112-113tr, Paul Hilton / Greenpeace 112bl, Egidio Trainito / Greenpeace 112br, Claudia Carrillo / Greenpeace 113bc
Born Free: ©Daily Mail 6tr, © alwynphoto.com 188-189tr, © Robin Claydon 188bl, © georgelogan.co.uk 188bc, © Sangha Pangolin Project/Maja Gudehus 188br, © BFF 189bc
World Land Trust: Callie Broaddus 6br, Robin Moore 204-205tr, Jo Dale 204bl, FUNDAECO 204br, Dan Bradbury / World Land Trust 205bc
Re:wild: Robin Moore / Re:wild 214-215tr, Aussie Ark 214bl, William Robichaud 214bc, Tilo Nadler 214br, David Stowe / Aussie Ark 215bc
Reserva: The Youth Land Trust: Highland Yearbook Staff 6cr, Callie Broaddus 218-219tr, Callie Broaddus 218bl, Sean Graesser 218bc, Callie Broaddus 218br, Callie Broaddus 219bc
Rewilding Europe: Robin Moore 6ctl, Sandra Bartocha / Wild Wonders of Europe 230-231tr, Staffan Widstrand / Rewilding Europe 230bl, Magnus Elancer / Wild Wonders of Europe 230bc, Mark Hamblin / Wild Wonders of Europe 230br, Fabrizio Cordischi / Rewilding Apennines 231bc
The Wildlife Trusts: Neil Aldridge 248-249tr, Malcolm Brown 248bl, David Parkyn 248bc, Andy Rouse / 2020VISION 248br, Neil Aldridge 249bc
Stop Ecocide International: Ruth Davey
look-again.org 6cbl
Callie Broaddus 7c, 8c, 9c
Dreamstime.com: André Costa 84-85
Native Daily Network 99t
Liberia Chimpanzee Rescue & Protection 173br
Brooklyn Beckham 294t, 294bl, 294br

Additional thanks:

p.6 with thanks to Hector Trunnec for artwork
pp.254-257 A Fevered Mind and a Pot of Pee: An image has been designed using resources from Freepik.com

DK would also like to thank Mark Penfound, Anne Damerell, Jessica DeFerry, Lottie Chesterman, Martin Way, Lisa Moore, Nicola Torode, and Tom Morse for their assistance on this title.

Editorial Beth Davies, Nicole Reynolds
Design James McKeag, Jennifer Murray
Senior Production Editor Jennifer Murray
Senior Production Controller Lloyd Robertson
Managing Editors Pete Jorgensen, Paula Regan
Managing Art Editor Jo Connor
Publishing Director Mark Searle

First American Edition, 2021
Published in the United States by DK Publishing
1450 Broadway, Suite 801, New York, NY 10018

21 22 23 24 25 10 9 8 7 6 5 4 3 2 1
001-324740-Nov/2021

Published in Great Britain by Dorling Kindersley Limited

A catalog record for this book is available from the Library of Congress.
ISBN 978-0-7440-4282-5

Printed and bound in China

For the curious
www.dk.com

This book has been produced using Forest Stewardship Council™ (FSC™) certified materials. FSC is the gold standard in forest certification schemes promoting responsible forest management and protecting the rights of indigenous populations. The inks have been manufactured using vegetable oils (rather than mineral oil). The cover is finished with a water-based varnish rather than a plastic laminate. Our print manufacturer is ISO 14001 accredited–this is an international standard for environmental management ensuring they are measuring their environmental impact and continually improving their performance year on year.